Safer Systems: The N

thescsc.org

scsc.uk

Related Titles

Assuring the Safety of Systems
Proceedings of the Twenty-first Safety-critical Systems Symposium, Bristol, UK, 2013
Dale and Anderson (Eds)
978-1481018647

Addressing Systems Safety Challenges
Proceedings of the Twenty-second Safety-critical Systems Symposium, Brighton, UK, 2014
Dale and Anderson (Eds)
978-1491263648

Engineering Systems for Safety
Proceedings of the Twenty-third Safety-critical Systems Symposium, Bristol, UK, 2015
Parsons and Anderson (Eds)
978-1505689082

Developing Safe Systems
Proceedings of the Twenty-fourth Safety-critical Systems Symposium, Brighton, UK, 2016
Parsons and Anderson (Eds)
978-1519420077

Developments in System Safety Engineering
Proceedings of the Twenty-fifth Safety-critical Systems Symposium, Bristol, UK, 2017
Parsons and Kelly (Eds)
978-1540796288

Evolution of System Safety
Proceedings of the Twenty-sixth Safety-critical Systems Symposium, York, UK, 2018
Parsons and Kelly (Eds)
978-1979733618

Engineering Safe Autonomy
Proceedings of the twenty-seventh Safety-Critical Systems Symposium, Bristol, UK, 2019,
SCSC-150
Parsons and Kelly (Eds)
978-1729361764

Assuring Safe Autonomy
Proceedings of the twenty-eighth Safety-Critical Systems Symposium, York, UK, 2020
SCSC-154
Parsons and Nicholson (Eds)
978-1713305668

Systems and Covid-19
Proceedings of the twenty-ninth Safety-Critical Systems Symposium, Online, 2021
SCSC-161
Parsons and Nicholson (Eds)
979-8588665049

Mike Parsons • Mark Nicholson
Editors

Safer Systems: The Next 30 Years

Proceedings of the 30th
Safety-Critical Systems Symposium
(SSS'22)
8th-10th February 2022

SCSC-170

thescsc.org scsc.uk

Editors

Mike Parsons and Mark Nicholson
SCSC & Department of Computer Science
University of York
Deramore Lane, York
YO10 5NG
United Kingdom

While the authors and the publishers have used reasonable endeavours to ensure that the information and guidance given in this work is correct, all parties must rely on their own skill and judgement when making use of this work and obtain professional or specialist advice before taking, or refraining from, any action on the basis of the content of this work. Neither the authors nor the publishers make any representations or warranties of any kind, express or implied, about the completeness, accuracy, reliability, suitability or availability with respect to such information and guidance for any purpose, and they will not be liable for any loss or damage including without limitation, indirect or consequential loss or damage, or any loss or damage whatsoever (including as a result of negligence) arising out of, or in connection with, the use of this work. The views and opinions expressed in this publication are those of the authors and do not necessarily reflect those of their employers, the SCSC or other organisations.

ISBN: 9798778289932

Published by the Safety-Critical Systems Club 2022.

Individual chapters © as shown on respective first pages in footnote.

All other text © Safety-Critical Systems Club C.I.C.

Cover design by Alex King

scsc.uk/scsc-170

Hardcopy available on Amazon

thescsc.org   scsc.uk

Preface

Our world has changed dramatically over the last two years with the Covid-19 pandemic, and life may only slowly be returning to normal. However, our 30th symposium brought some new opportunities, new thinking and also new delegates. The symposium had a special flavour, celebrating the club's successes over the years and giving a taste of what the next 30 years in safety engineering might be like.

The management of the pandemic has used safety systems like never before: everything from new variant modelling through building ventilation analyses to vaccination passports. Your smartphone likely ran safety-related functions linked to the virus. Many of these solutions were not recognized as safety-related and assurance in our normal sense was often missing. There was no doubt that lives were lost by failures of these systems.

Our established engineering-led solutions continued development throughout the year. Autonomy in road vehicles became ever closer with functions like automatic lane-keeping; use of AI increased and continued to be deployed in everything from credit scoring to interpretation of medical images.

The SCSC Working Groups have evolved to cover areas such as Multicore and Manycore processor-based systems, Safety Culture, and underlying Risk Ontologies; the existing groups for Assurance Cases, Autonomous Systems, Data Safety, Security Informed Safety and Service Assurance have all continued throughout the year.

This forward-looking symposium tackled many of these new challenges and presented work towards practical and industrially relevant solutions. The primary themes were: Assurance, Autonomy, Ethics, Human Factors, New Techniques, New Applications, Healthcare Systems and Safety and Security

SSS'22 ran across three days of presentations, grouped into themes. The keynote speakers were: Graham Braithwaite, Dewi Daniels, Adam Johns, Peter Ladkin, John McDermid, Catherine Menon, Reuben McDonald, Paula Palade and Harold Thimbleby. Wendy Owen gave a talk looking back at 30 years of the Newsletter; Tom Anderson gave the after-dinner talk on the Wednesday and Tim Kelly gave a thoughtful closing talk.

We are very grateful to all those who helped organise the event within the SCSC with the support of the University of York.

Mike Parsons and Mark Nicholson

SCSC Mission and Aims

Mission

To promote practical systems approaches to safety for technological solutions in the real world.

Where *"systems approaches"* is the application of analysis tools, models and methods which consider the whole system and its components; *"system"* means the whole socio-technical system in which the solution operates, including organisational culture, structure and governance, and *"technological solutions"* includes products, systems and services and combinations thereof.

Aims

1. **To build and foster an active and inclusive community of safety stakeholders:**
   a. "safety stakeholders" include practitioners (in safety specialisms and other disciplines involved in the whole lifecycle of safety related systems), managers, researchers, and those involved in governance (including policy makers, law makers, regulators and auditors)
   b. from across industry sectors, including new and non-traditional areas
   c. recognising the importance of including and nurturing early career practitioners
   d. working to remove barriers to inclusion in the community
2. **To support sharing of systems approaches to safety:**
   a. enabling wider application
   b. supporting continuing professional development
   c. encouraging interaction between early career and experienced practitioners
   d. using a variety of communication media and techniques to maximise coverage
   e. highlighting the lessons which can be learned from past experience
3. **To produce consistent guidance for safety stakeholders where not already available.**
   a. "consistent" meaning the guidance is consistent within itself, and with other guidance provided by SCSC; although SCSC will also aim to co-ordinate with external guidance this is more difficult to achieve
4. **To influence relevant standards, guidance and other publications.**
5. **To work with relevant organisations to provide a co-ordinated approach to system safety.**
6. **To minimise our environmental impact wherever possible.**

The Safety-Critical Systems Club

organiser of the

Safety-Critical Systems Symposium

Avoiding Systems-Related Harm

The past two years have been heavily impacted by Covid-19 with things only slowly returning to some sort of normality. Many systems have been involved with the pandemic management, and the SCSC has been tracking issues with these systems. This is in accordance with the mission of the SCSC: using systems approaches to prevent harm.

Many safety-critical systems in everyday life work as expected day-in day-out. Safety of these systems is accepted as routine: airbags in vehicles, air-traffic control, vaccine passport apps and infusion pumps are some of the critical systems in use, on which life and property depend.

That safety-critical systems and services do work is because of the expertise and diligence of professional engineers, regulators, auditors and other practitioners. Their efforts prevent untold deaths and injuries every year. The Safety-Critical Systems Club (SCSC) has been actively engaged for 30 years to help to ensure that this is the case, and to provide a "home" for safety professionals.

What does the SCSC do?

The SCSC maintains a website (thescsc.org, scsc.uk), which includes a diary of events, working group areas and club publications. It produces a regular newsletter, *Safety Systems*, three times a year and now also a peer-reviewed journal published twice a year. It organises seminars, workshops and training on general safety matters or specific subjects of current concern.

Since 1993 it has organised the annual Safety-Critical Systems Symposium (SSS) where leaders in different aspects of safety from different industries, including consultants, regulators and academics, meet to exchange information and experience, with the content published in this proceedings volume.

The SCSC supports industry working groups. Currently there are active groups covering the areas of: Assurance Cases, Autonomous Systems, Data Safety, Multicore and Manycore, Ontology, Security Informed Safety, Service Assurance and Safety Culture. These working groups provide a focus for discussions within industry and produce new guidance materials.

The SCSC carries out all these activities to support its mission:

... To promote practical systems approaches to safety for technological solutions in the real world....

Origins

The SCSC began its work in 1991, supported by the UK Department of Trade and Industry and the Engineering and Physical Sciences Research Council. The Club has been self-sufficient since 1994. In 2021 it became a separate legal entity.

Membership

Membership may be either corporate or individual. Membership gives full web site access, the hardcopy newsletter, other mailings, and discounted entry to seminars, workshops and the annual Symposium. Membership is often paid by employers.

Corporate membership is for organisations that would like several employees to take advantage of the benefits of the SCSC. Different arrangements and packages are available. Contact alex.king@scsc.uk for more details.

New is a short-term Publications Pass which, at very low cost, gives a month's access to all SCSC publications for non-members.

More information can be obtained at: scsc.uk/membership

Club Positions

The current and previous holders of the club positions are as follows (past holder in *italics*):

Leader/Director
Mike Parsons   2019-
Tim Kelly   *2016-2019*
Tom Anderson   *1991-2016*

Newsletter Editor
Paul Hampton   2019-
Katrina Attwood   *2016-2019*
Felix Redmill   *1991-2016*

eJournal Editor
John Spriggs   2021-

Website Editor
Brian Jepson   2004-

Steering Group Chair
Roger Rivett   2019-
Jane Fenn (deputy)   2019-
Graham Jolliffe   *2014-2019*
Brian Jepson   *2007-2014*
Bob Malcolm   *1991-2007*

University of York Coordinator
Mark Nicholson   2019-

Coordinator/Events Coordinator/Programme Coordinator
Mike Parsons   2014-
Chris Dale   *2008-2014*
Felix Redmill   *1991-2008*

Safety Futures Initiative Lead
Zoe Garstang   2020-
Nikita Johnson   *2020-2021*

Manager
Alex King   2019-

Administrator
Alex King   2016-
Joan Atkinson   1991-2016

Current Working Group Leaders
Assurance Cases	Phil Williams
Security Informed Safety	Stephen Bull
Data Safety Initiative	Mike Parsons
Autonomous Systems	Philippa Ryan
Multicore and Manycore	Louise Harney/Lee Jacques
Safety Culture	Michael Wright
Risk Ontology	Dave Banham
Covid-19	Peter Ladkin
Service Assurance	Mike Parsons

SCSC Organisation

The SCSC is a "Community Interest Company" (CIC). A CIC is a special type of limited company which exists to benefit a community rather than private shareholders. The SCSC has:

- A 'community interest statement', explaining our plans;
- An 'asset lock'- a legal promise stating that our assets will only be used for our social objectives;
- A constitution, and
- Approval by the Regulator of Community Interest Companies.

Our community is that of Safety Practitioners. As a distinct legal entity the SCSC has more freedom and can legitimately do things such as make agreements with other bodies and own copyright on documents.

There is no change to the way we do things, or the membership we serve.

Our Company Number is 13084663.

Contents

(In presentation order; see also full Author Index *at page 327)*

Keynote:
Consumerism, Contradictions, Counterfactuals: Shaping the Evolution of Safety Engineering
John A McDermid, Zoë Porter, Yan Jia 15

Does an agile approach improve the way we derive RAMS requirements?
Gavin Wilsher 37

Design-time Specification of Dynamic Modular Safety Cases in Support of Run-Time Safety Assessment
Elham Mirzaei, Carmen Cârlan, Carsten Thomas, Barbara Gallina 51

Keynote:
Stories and narratives in safety engineering
Catherine Menon, Austen Rainer 75

Development of Rechargeable Electrical Energy Storage Systems for Automotive and Aviation
Paul Malcolm Darnell, Pavan Venkatesh Kumar 87

Keynote:
Ethics and Safety for Connected and Automated Vehicles
Paula Palade 115

Towards a robust safety assurance process for maritime autonomous surface ships
Gerasimos Theotokatos, Victor Bolbot, Evangelos Boulougouris, Dracos Vassalos 121

A Step-by-Step Methodology for Applying Service Assurance
James Catmur, Mike Parsons, Mike Sleath 123

Keynote:
The German and Belgian Floods in July 2021
Peter Bernard Ladkin 149

Keynote:
Introducing a Restorative Just Culture and the Learning Review at the Docklands Light Railway
Adam Johns 181

Human Reliability in Complex Systems
Rachel Selfe 183

Keynote:
Thirty years of learning by accident
Graham Braithwaite 201

Keynote:
"Cowboy digital" undermines safety-critical systems
Harold Thimbleby 203

Could the Introduction of Assured Autonomy Change Accident Outcomes?
Dewi Daniels, Chris Hobbs, John McDermid, Mike Parsons, Bernard Twomey
.... 227

A Pipeline of Problems, or Software Development Nirvana?
The Challenges of Adopting DevSecOps in a Safety-Critical Environment
Mike Drennan, Paul McKernan, James Sharp 263

Formal verification of railway interlocking and its safety case
Alexei Iliasov, Dominic Taylor, Systra Scott Lister, Linas Laibinis, Alexander Romanovsky 281

Safety-critical Multi-core for Avionics
Gary Gilliland 299

Keynote:
At the interface of engineering safety and cyber security
Reuben McDonald 313

Keynote:
What do Byzantine Generals and Airbus Airliners Have in Common?
Dewi Daniels 315

AUTHOR INDEX **327**

Consumerism, Contradictions, Counterfactuals: Shaping the Evolution of Safety Engineering

John A McDermid, Zoë Porter, Yan Jia

Assuring Autonomy International Programme, University of York

Abstract *This paper takes the (perhaps unusual) view that consumerism has helped to drive improvements in safety over the years. However, the successes in terms of the availability of (safe) goods and services, e.g. cars and cheap air transport, present contradictions (or ironies) in terms of the subsequent impact on the environment which ultimately has a deleterious effect on safety and well-being. These contradictions suggest the need to re-frame safety engineering. The paper proposes an approach based on the notion of well-being and discusses how counterfactuals might play a role in analysing and communicating about safety concerns.*

1 Introduction

In several domains, sustained improvements to system safety have been achieved over many decades. This is perhaps particularly obvious in road vehicles and air transport, but the trend can be seen more generally. Some of the credit for this is due to good safety engineering and safety culture. But customer pressure is also a factor as safety has become essential to the sales of some products or services. This combines with other factors, such as progressive reductions in unit prices, to enable widespread availability of safe products and services. But it also comes at a cost. The most obvious is the environmental impact of increased road traffic and air travel which, in turn, has a negative impact on human safety. Because of ironies or contradictions such as these – whereby improvements to safety lead indirectly to greater physical risk – we propose a re-framing of safety engineering and assurance to look much more broadly and holistically at hazards and impacts. We also propose a re-framing of safety engineering and assurance to include (and codify) the intended benefit from the system as well as the possible harm. This widening of the safety and assurance landscape aligns with the precepts of 'Ethically Aligned Design' (IEEE 2019), and the principles of Responsible Research and Innovation (Owen et al. 2013; Von Schomberg 2013), but we seek here to

© J A McDermid, Z Porter, Y Jia 2022.
Published by the Safety-Critical Systems Club. All Rights Reserved

define a more tangible approach to defining and assessing risk, more strictly a balance of risk and benefit.

Aligned with the re-framing of safety as a concern we also consider the role of counterfactual reasoning in the evaluation, communication, and mitigation of risk. The term 'counterfactual' has a long history in philosophy and related disciplines. A counterfactual statement can be defined loosely as being about "what was not, or is not, but could or would have been" and it is often expressed as subjunctive conditional: "if x had/had not occurred, then y would/would not have occurred" (Starr 2021). Although 'counterfactual' is now used in artificial intelligence (AI) to refer to a particular form of explanation method, our focus here is rather different. We consider several potential roles for counterfactual reasoning in this re-framing of safety engineering, e.g. supporting accident analysis.

The rest of the paper is structured as follows. Section 2 considers the relationship between *consumerism* and historical trends towards safer products and services. Section 3 considers the *contradictions* or ironies that arise from the successful reduction in cost, and improvements in safety, of products and services, particularly those that are sold on the mass market. Section 4 proposes a progressive re-framing of safety engineering which embraces the notion of well-being, codifies benefit as well as harm, and considers longer-term impacts such as environmental damage. The intent is that this resonates more fully with ethical and societal concerns and expectations about high-integrity systems, and that these ideas will help to shape the evolution of safety as a concern, and safety engineering as a discipline, to give a practical basis for ethically aligned design and systems engineering. Within this, we explore the potential role of counterfactual analysis in understanding and communicating about hazards and risks. In Section 5, we discuss how the proposed re-framing identifies and addresses issues that are not covered in current enlargements of safety engineering, such as Safety II (Hollnagel 2018), and new ethical standards for engineers, although the ideas presented here are complementary to those developments. We also raise open and unanswered questions, such as formalisations of 'well-being' and distribution of risk, emphasising the importance of multidisciplinary research in this evolving concern. Finally, we consider what steps might need to be taken to enable these broad concepts to influence real-world engineering.

2 Consumerism and the Achievement of Safety

Mature industries, such as aerospace and automotive, have seen sustained reductions in accident levels and fatalities over many years albeit with some geographical variation. In our view this is *in part* due to good safety engineering and safety management, but there are also other influences, such as the (implied) pressures on manufacturers to achieve societally acceptable levels of risk, the cost of recalls and potential reputational damage, which drive improvements to safety. Here we highlight one influence that seems to us to be particularly significant – consumerism. Consumerism has two definitions, or facets, which we might characterise as:

- The protection or promotion of the interests of customers and
- The preoccupation of society with the acquisition of goods or services.

Whilst these might seem somewhat contradictory, these facets *work together* when it comes to the impact on safety. We illustrate this by considering the very different effects of consumerism in two sectors: air transport and cars.

2.1 Air Transport and Tourism

One of the most famous graphics showing how air travel has become safer over the years is from Boeing's annual aviation statistics summary (Boeing 2020). Figure 1 shows the data from the late 1950s onwards. From a safety engineering perspective, this indicates (although it doesn't prove) the long-term effectiveness of safety analysis and management (notwithstanding the issues surrounding the Boeing 737 MAX). In particular, the approach to analysing accidents has meant that the industry has understood the underlying causes of accidents, e.g. "power structures" in cockpits, and introduced specific remedies such as approaches to crew/cockpit resource management (CRM) to reduce problems of communication and losses of situational awareness.

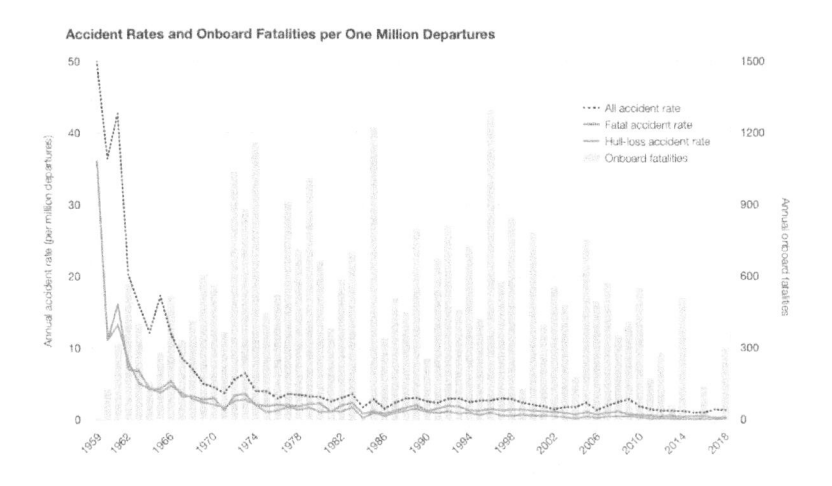

Fig. 1. Aircraft Accident Statistics (Boeing 2020)

Worldwide, aircraft departures have grown substantially over the years. They stood at over 35 million flights per annum in 2019, although this has dropped dramatically due to Covid-19 (World Bank 2021). Considering Figure 1, if the accident rate had remained at 1960s levels, then theoretically there would be roughly one aircraft accident per day. Of course, this is just hypothetical – given the impact of such accidents on public perception, and the impact of court cases and compensation claims, such an accident rate would very likely cause aircraft usage to drop dramatically, indeed to fundamentally transform the air transport sector.

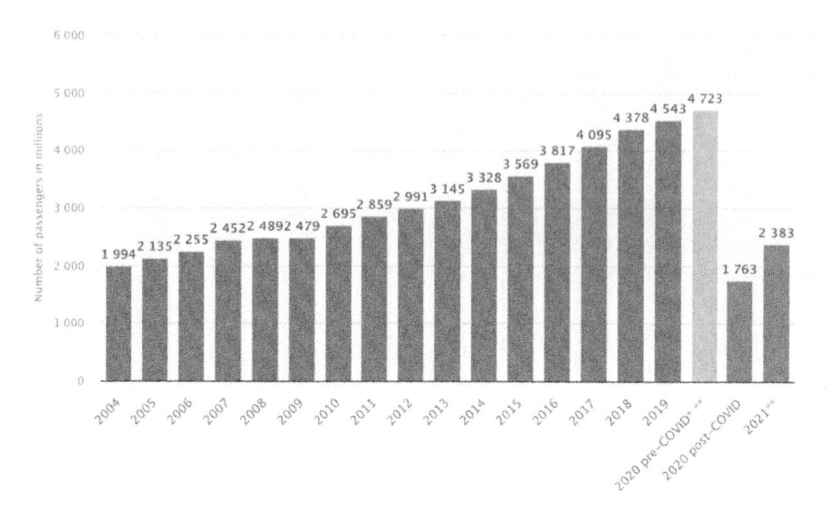

Fig. 2. Aircraft Passenger Numbers (Statista 2021)

So, what has driven this growth in traffic? The answer appears to be tourism which, of course, is an example of consumerism where people are interested in acquiring services – in this case cheap (international) holidays – resulting in the growth in passenger numbers illustrated in Figure 2. Air travel has changed from being a privilege to being commonplace, and an analysis suggests that tourism had become the world's largest industry by 1984 (Lyth and Dierikx 1994) and it stood at over 10% of *global* Gross Domestic Product (GDP), pre-pandemic.

Whether or not aircraft safety would have improved so much without the pressures of consumerism is, of course, unknowable; we cannot show cause and effect. However, the counterfactual is clear – without the safety improvements, the traffic growth and scale of international tourism would not have occurred as society would not have embraced air travel if the apparent risks of flying were so high.

2.2 Cars and Shifting Expectations

The data from the automotive sector also show a significant downward trend in accidents over the years. Figure 3 is from a Department for Transport (DfT) summary for Great Britain (GB) over a 40-year period (Department for Transport 2020). As with air transport, this downward trend is against an increase in traffic volumes – with a fatal accident every 4.9 billion vehicle miles in 2019 against one every 7.1 billion miles in 2009. However, what is perhaps more telling is the improvements in vehicles that have occurred over many decades.

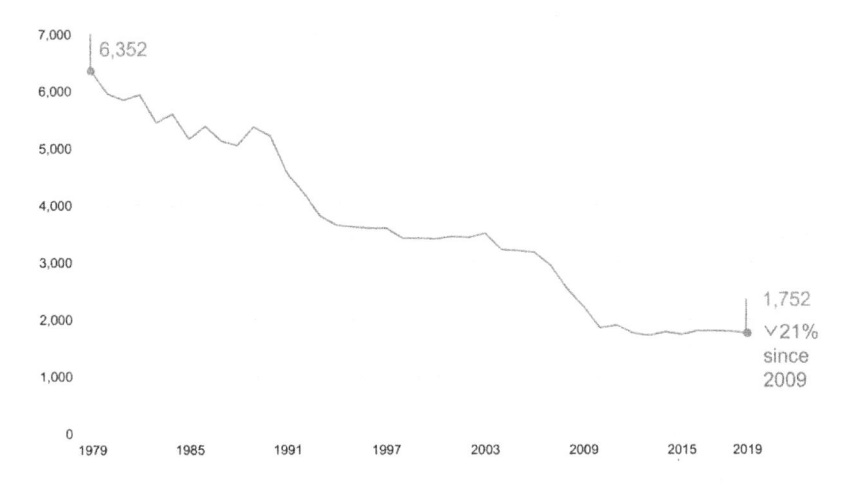

Fig. 3. Fatalities in Reported Road Accidents, GB 1979-2019 (DfT 2020)

First, it is worthwhile making some observations about consumerism. To be successful, products often need differentiators, or unique selling points (USPs), that set them apart from the competition. There are also minimum expectations on products and if these are not met by a particular product then it might not be successful, despite the presence of some attractive USPs. These minimum expectations (sometimes referred to as "table stakes" based on the use of the term in gambling) can change over time – indeed something that was once a USP can become a "table stakes" feature. This can be seen for cars, including for safety features, which is illustrated by the following partial timeline extracted from one produced by the UK Automobile Association (AA 2021)):

1911 – rear view mirrors.
1921 – headrests.
1951 – airbags.
1952 – crumple zones.
1963 – inertia reel seatbelts.

1978 – anti-lock braking.
1991 – side impact protection systems.
1995 – electronic stability control.
2000 – lane departure warning systems.
2008 – autonomous emergency braking.
2010 – pedestrian detection system.

Many of these early innovations have become expected and some, e.g. seat belts, are now required by regulation. There is also a significant shift from "passive" safety, e.g. headrests, through to more "active" safety systems such as autonomous emergency braking. One can also add many items to this list that were once seen as "luxuries", e.g. reversing cameras, which are now fitted on many vehicles.

Despite a similar effect on safety, the trends and impact of consumerism in the automotive sector is very different from than in air transport. Aircraft purchase (or lease) is the province of the professional and consumer pressure for safe, cheap services is indirect. Cars are (often) an individual purchase and the availability of safety features on some vehicles pushes the manufacturers to provide similar capabilities for fear of losing sales to other brands. Also, customer attitudes are important – it is not uncommon for people to say: "I won't buy it unless it has X" (where X is some safety feature, e.g. anti-lock brakes). Finally, the (European) New Car Assessment Programme (Euro NCAP)[1] serves to keep safety in the public eye and a factor when purchasing a new vehicle. From a safety perspective this is good news as vehicle manufacturers strive to improve their NCAP rating, and there are now trends to protect vulnerable road users (VRUs).

There is another perspective on consumerism in the automotive sector which is relevant to our re-framing of safety. Electric cars first appeared as early as the 1830s, and were quite widespread by 1900, particularly in cities with good availability of electricity supplies (Department of Energy 2014). At the same time, cars based on the internal combustion engine were dirty and smelly – presaging today's problems – and had other challenges, including being hard to drive. But the availability of cheap fuel and cheaper cars (a Model T Ford was about a third of the price of a similar electric vehicle) meant that more consumers could acquire goods by buying vehicles with internal combustion engines. Figures 4 and 5 are examples of advertisements that reflect the difference in price of electric vehicles and those with internal combustion engines; although the adverts aren't strictly contemporaneous the difference in price ($2,250 vs $360) is indicative of the competitive problems of electric vehicles. Consequently, electric vehicles had almost disappeared by the 1930s. Again, a plausible counterfactual can be made – without the availability of cheap cars using internal combustion engines and cheap fuel, city transport would not have moved away from electric vehicles (see

---

[1] See: https://www.euroncap.com/en

section 3.2 for a discussion of the wider implications of this change in propulsive power).

Fig. 4. Electric Car advert circa 1900

Fig. 5. Model T Ford advert circa 1925

3 Contradictions and the Wider Impact on Society

The positive impacts of consumerism are offset by some negative effects – which in this paper we call contradictions (although strictly they are more 'ironies' than 'contradictions'). Our concern here is primarily those contradictions related to safety. Consumerism has led to improvements in safety, considering the products or services in themselves, but with negative safety effects if we look more widely. We focus on two environmental impacts – global warming and air quality in cities.

3.1 Global Warming

There is growing evidence that human activity is a major contributor to global warming. There are many factors, including the generation of electricity from fossil fuels, but our focus is transport. The International Panel on Climate Change (IPCC) recently stated 'carbon dioxide (CO_2) is the main driver of climate change, even as other greenhouse gases and air pollutants also affect the climate'; tourism contributes about 8% of emitted CO_2 and air travel is about half of that (IPCC 2021). Whilst the data is not so conclusive, there are negative effects of emission at altitude, e.g. NO_2 from aircraft. Road transport contributes significantly to CO_2 emissions with around 10% from freight and 15% from passenger transport (Ritchie 2021).[2]

The impact of global warming on human health and safety is evident from the growing frequency of extreme events, including recent destructive hurricanes which have affected the USA. Furthermore, predictions of sea level rise suggest that at least 200 million people will be living below sea level by the end of the century and as many as 630 million are projected to live below annual flood levels on a high emissions scenario (Kulp and Strauss 2019). Alarming as they are, these predictions do not include the impact of melting of the ice shelf.

Thus, although aircraft and cars are remarkably safe in themselves, and becoming more so due to the forces of consumerism, amongst other things, they both contribute to global warming, which has the potential to be an existential threat to humans and many other species.

3.2 Air Quality in Cities

Emissions from road vehicles have an impact on air quality, especially in cities, which, in turn, has an impact on human health. The cause-and-effect relationships between pollution and health are complex to establish. The Committee on the Medical Effects of Air Pollutants points out the dissenting views in the committee on these causal questions (COMEAP 2018).

The COMEAP report presents a detailed analysis of the effect of NO_2 and particulate matter (PM) on premature mortality, seeking to adjust for the correlations between NO_2 and PM, as both can be produced in vehicle emissions. The report contains the estimate that premature deaths in the UK in 2013 were in the range 22,000 to 36,000. The report aims to be balanced and includes questions from dissenters about the (evidence for) causal relationships between NO_2 and premature mortality, however the authors all agree that a reduction of NO_2 and PM will be beneficial to health. The report also considers counterfactuals – in this case, baseline levels of NO_2 and PM against which measured pollution is compared. There are some detailed studies of the effects of PM that identify significant levels of premature mortality in international cohort studies (Burnett et al 2018). Further, there has recently been a conclusion by a coroner that emissions contributed to the death of a schoolgirl in London (The Sunday Times 2020).

As with climate change there is a contrast, or contradiction, between the safety of road vehicles, in themselves, and the wider and longer-term impact of their use. The 1,752 fatalities on UK roads in 2019 due to vehicle accidents is less than

---

[2] N.B. The figures from (Ritchie 2021) are scaled to be consistent with the IPCC report, but there is not an exact match so the figures should be taken to be indicative not absolute

a tenth of the estimated 22,000 to 36,000 premature fatalities due to poor air quality. As noted above, if electric vehicles had remained a major form of transport in cities as they were over a century ago, then this impact would have been much reduced.

4 Re-framing System Safety Engineering

To resolve contradictions such as the indirect negative safety impact from environmental damage of heavy road and air traffic, we need to broaden safety engineering beyond the normal (narrowly defined) consequences of hazards. Further, we need to include benefits as well as harms and to consider trade-offs between potentially incommensurate benefits and harms. We build up to the re-framing of safety engineering and assurance progressively in the rest of this section.

4.1 The Trade Space

The first step is to define what we call the "trade space" – the range of anticipated impacts that will need to be included in any risk assessment and when evaluating trade-offs. There is an obvious trade-off between the two aspects of consumerism – having fewer cars, aircraft, etc. reduces availability of goods and services, but promotes the interests of consumers in terms of reducing the risks to health and safety arising from adverse environmental impacts. Staying with cars as an example, reducing their availability might mean that an individual travels more by bicycle – at one level, this is good for their health, but at another level they become a VRU and are at about 25 times greater risk per mile travelled than car occupants (Department for Transport 2020).

There will be many other benefits and harms associated with a given product or service – for the owner/user, for the designer or manufacturer, for people directly affected by the system, e.g. those in a city where a car is used, and for society in general. For example, using electric vehicles in a city is beneficial in terms of pollution and hence air quality – but this may just displace pollution rather than reduce it, depending on how the electricity is generated. If it is produced using fossil fuels, then there may be a similar amount of pollution, just in a different place. There are, of course, ethical questions about the acceptability of actions which shift risks (and benefits) between different groups. We identify this as an open question (or future work) in Section 5. It is also worth noting that there is a substantial environmental impact from making cars; whilst there are varying analyses, some suggest that manufacturing and driving a car have similar carbon footprints (Berners-Lee and Clark 2010), and manufacturing electric vehicles has a greater impact than making a comparable conventional vehicle. However, those employed at the factories gain benefits as well as being exposed to the localised risks so, again, there are trade-offs.

In practice, many of these factors are incommensurable. For example, the benefits of being employed at a car factory and the quality of life arising from paid work (psychological, societal), potential harms from the manufacturing processes (physical), and long-term impact from environmental damage cannot obviously be measured or evaluated on a single scale. In addition, given the nature of the supply chain for cars, the risks are quite widely distributed – in mines and quarries, in electronics factories, and so on – making it very hard to calculate risks and to undertake systematic risk-benefit trade-offs. Thus, the first issue is how to re-frame safety engineering to provide a "trade-space" which can be thought of as providing a framework of the different factors – benefits and harms – that need to be considered. We approach this from the viewpoint of well-being.

4.2 Focusing Safety Engineering on 'Well-Being'

The concept of 'dependability' has been long-used to embrace failure-related system properties – safety, availability, reliability, etc. (Avizienis 2004). The models underlying this concept assist with reasoning about the relationship of these key system properties, but dependability does not cover the wider impact on society considered in section 3, and the enlarged trade space described above. The concept of dependability is too narrow, but what is a suitable alternative? We believe that a human-focused approach is essential. Safety is about protecting people from harm. But in the face of contradictions and ironies of long-term negative effects on safety and human well-being from products and services that are "safe in themselves", it seems clear that safety engineering needs to evolve, and to be re-framed to consider not just individual but also societal and environmental impacts.

In philosophy, 'well-being' is what makes life good for the individual living that life – or how well a person's life is going for them (Crisp 2021). It is common for philosophers to draw a distinction between subjective and objective conceptions of well-being or welfare: broadly, whether the concept should be understood in terms of people's own preferences and accounts of what makes life good for them, or in terms of what objectively makes their life go well for them irrespective of their personal predilections. Hybrid theories combine objective and subjective elements of well-being (Parfit 1984). We seek to abstract the following discussion from a commitment to a specific theory of well-being. But any model of well-being that is applied in a re-framing of safety engineering will need to be to some degree objective and codified. The aim is for rational, repeatable safety engineering processes.

One policy approach to well-being with philosophical roots is the capabilities approach – that people need certain capabilities to function well; this derives from Aristotle's notion of *eudaimonia*, or flourishing, as the goal for humans (Nussbaum and Sen 1993). Developing this perspective, many policy-focused analyses decompose the notion of well-being. Some do so on the basis of needs, with, for example, health and personal autonomy taken to be primary, supported by secondary attributes such as nutritional food and clean water, adequate housing, a safe work environment, physical security, and so on (Doyal and Gough 2011).

Another interesting perspective is from Buddhist economics (Schumaker 1966). This considers wider impacts following the use of a product or service, investigating how trends affect individuals, society, and the environment, and links particularly well to the concerns introduced in section 3.

But how can we use the concept of well-being as a basis for enlarged analysis of system safety? We propose a two-layer model. The top level would consider, for a given or proposed system, the potential benefits and harms to individual well-being, society, and the environment. The identified concerns at this level would scope more detailed, lower-level analysis, for example informed by secondary attributes (Doyal and Gough 2011), for identifying benefits and harms (forms of hazard) in sufficient detail so trade-offs and tensions can be considered, and controls defined.

The top level is captured in Table 1. It focuses on impacts at the system level.

Table 1: Categories of Benefit and Harm

Benefits	Hazards or Harms
Individual/personal - Physical - Psychological - …	Individual/personal - Physical - Psychological - …
Societal	Societal
Environmental	Environmental

Physical impacts on individuals include improvements to physical safety as well as loss of life or bodily injuries. Psychological impacts include benefits to mental health, and hazards such as addiction and trauma. Societal impacts include benefits and harms to infrastructure and societal functioning (Hassel and Cedegren 2021). Societal impacts also arise from changes to risk distributions. Environmental impacts include issues such as air and water quality and it might be argued that loss of biodiversity has an impact on psychological well-being. In addition, how widely deployed the system is has societal and environmental implications (see 4.3 below). The scope of impact, which we consider under the broad term 'well-being', enlarges safety engineering, both as a discipline and a concern.

4.3 The Numbers Game

After defining the "trade space' and refocusing safety engineering on an enlarged conception of well-being, it remains to consider the impacts and hazards of widely deployed systems beyond the immediate and discrete impact on individuals. This ties into societal and environmental concerns. Safety engineering normally focuses on a single product or system. By way of illustration, we consider aviation. Safety targets are typically related to hazard classes, e.g. an occurrence rate of $< 10^{-9}$ per flight hour for catastrophic hazards. Such targets apply whether there are only a few aircraft of the type, e.g. Concorde, or a very widely deployed system, e.g. Boeing 737s. When we consider environmental hazards, aircraft-for-aircraft, Concorde would have had a greater environmental impact than an individual Boeing 737 or an Airbus 320. But since there are around 5,000 each of the 737 and 320 in service, their cumulative impact is much greater. In the early days of aviation, environmental impacts were a relatively minor concern. There were very small numbers of aircraft and accident rates were high, so a focus on the direct hazards to occupants made sense. As the analysis in Section 3.2 shows, this is no longer the case, and the sheer volume of air traffic contributes to global warming and thus poses an (indirect) safety risk. So, the next step is that some of the harms (and benefits) need to be considered for whole fleets, not just for individual systems.

Fig. 6. Concorde Leaving New York

4.4 The Interconnectedness of Benefits and Harms

It is also necessary to understand the dependencies and relationships between the different benefits, hazards, and concerns. Over time, regulations have been introduced to address environmental impacts of aircraft (including noise as well as emissions) but again these tend to be at the level of individual aircraft and are disjoint from other safety requirements. Initiatives such as "Net Zero"[3] take a more holistic approach to managing emissions, but not integrating different perspectives such as flight safety with environmental impact. This lack of integration makes it hard to balance different safety concerns. Therefore, it is necessary to consider the dependencies between the different safety concerns to manage them effectively, including indirect hazards to safety from other impacts, such as environmental damage. This consideration will include making trade-offs between both direct and indirect risk related to the same kind of hazard (e.g. to the physical safety of individuals) as well trade-offs between different kinds of hazard (e.g. safety and privacy).

4.5 Safety Engineering Re-framed: Motivational Examples

Safety processes normally start with Hazard and Risk Analysis (HARA). To take this broader view of safety (re-framing it) we suggest a precursor analysis using the notion of benefits and harms to individuals, society, and the environment to scope the issues to be addressed in HARA. We illustrate this by means of two examples.

The CHIRON project is developing a social care robot (see Figure 7), intended to help the elderly and infirm to stand, and thus to continue living independently. The safety of this system has been investigated with funding from the Assuring Autonomy International Programme (AAIP)[4] and this work identified some concerns beyond classical safety issues. These are identified (and amplified) below:

1) Individual
    a) Benefit – enhanced/continued independence (psychological).
    b) Harm – injury from fall (physical); loss of strength/capability over time due to system providing excessive assistance (physical); reduced mental health due to isolation (psychological).
2) Society
    a) Benefit – reduced demand on social care system.

---

[3] See: https://www.gov.uk/government/publications/net-zero-strategy

[4] See: https://www.york.ac.uk/assuring-autonomy/projects/assistive-robots-healthcare/

b) Harm – growth in numbers of isolated elderly/infirm individuals later requiring mental health or other support.
3) Environment
   a) Benefit – minimal.
   b) Harm – minimal.

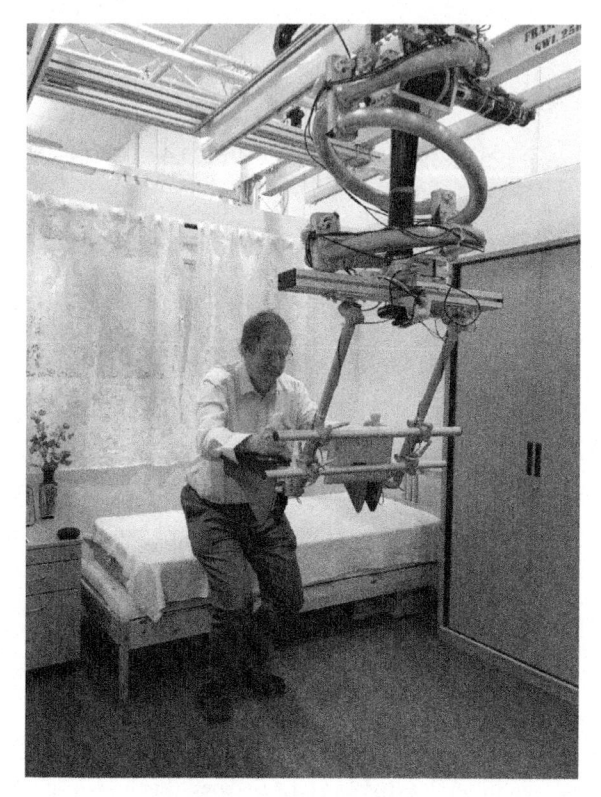

Fig. 7. The CHIRON robot

Traditional safety engineering would address injury from falls, but the other issues require multidisciplinary input, e.g. from physiology and sociology. Broader models, e.g. of the social care system, are also needed for a complete analysis. There will potentially be environmental impact from developing the system, but this is assumed to be minimal as the number of systems is likely to be limited (see the next example for a discussion of supply chain impacts).

The UK has committed to phasing out (new) petrol and diesel cars by 2030. To give more focus, and noting the fact that safety analysis normally addresses particular products or services, we consider delivery vehicles (e.g. developed by

Arrival)[5] but without autonomy, i.e. we assume that the vehicles have a human driver. Here the primary individuals affected are the delivery drivers and those working in factories producing the vehicles:

1) Individual
   a) Benefit – improved air quality for drivers (physical); reduced exposure to hazards from factory automation[6] (physical).
   b) Harm – injury from battery fires (Chen et al 2021) (physical); injury from handling toxic materials in factories (physical).
2) Society
   a) Benefit – ability to deliver products to cities without adversely impacting air quality; enhanced employment and economic benefits for the UK (VividEconomics 2018).
   b) Harm – no obvious issues.
3) Environment
   a) Benefit – aluminium and thermoplastic construction means that the van is light (so there is less impact on the road) and it is more sustainable meaning less impact from re-manufacturing; potential for reduction in pollution long-term through recycling, etc.
   b) Harm – impact in the supply chain from mining for the materials needed for batteries, and their transportation; impact of disposal of batteries at the end of life (Kang et al 2013).

In environmental terms there are also potential impacts of shifts in where the energy is generated, see the discussion in section 3.2. Due to the nature of the system, a long-term view needs to be taken, e.g. of environmental impact and harm.

4.6 Safety Engineering Re-framed: Towards a Process

The AAIP is developing revised safety processes for autonomous systems and robots, including a phase referred to as Societal Acceptability of Autonomous Systems (SOCA). We see the sorts of issues raised here as informing how a broader range of ethical and societal impacts can be incorporated into the assurance framework, but SOCA will also need to address the transfer of decision-making responsibility from humans to machines – and the distinct challenge that autonomy brings.

---

[5] See: https://arrival.com/?topic=products
[6] See: https://www.wired.co.uk/article/arrival-electric-vehicles-microfactory

As indicated above, subsequent analyses need to be informed by this scoping. We envisage enhancements of classical analysis techniques, e.g. HAZOP, for such stages but discussion of such techniques is outside the scope of this paper.

It is common, in many industries, to produce assurance cases to support decisions to approve a system for use. If assurance cases continue to be used then they need to be expanded to address impacts on society and the environment, as well as to the individuals directly (and indirectly) affected by the system, perhaps ultimately taking a 'Global Safety' perspective. There will also be a need to reason about benefits vs harms (risks). In current practice, arguments of risk versus benefit, including considering costs of options, are carried out where there is difficulty in reaching risk targets and the developers wish to show that risks are reduced as low as reasonably practicable (ALARP). Given the re-framing of safety we propose here then the arguments of benefit versus harm needs to be considered in all cases, not just *in extremis*, to seek to identify an "optimal" system concept and implementation. Here, the "costs" will need to include harms to society and environment, and not just focus on engineering economics. In principle it is (legally) necessary to produce ALARP arguments comparing all possible designs. There is a major problem with ALARP arguments which becomes worse in the re-framed safety process – and here we see a role for counterfactual thinking.

4.7 Safety Engineering Re-framed: Counterfactuals through Life

In Section 2, we indicated how counterfactual thinking can contribute to reasoning about longer-term and indirect impacts to human well-being from high-integrity systems. Counterfactual explanations have been adopted by the AI and machine learning (ML) communities as a way of explaining the behaviour of otherwise opaque algorithms (Wachter et al 2018). They are a form of example-based reasoning (Miller 2019). They are likely to have a role in safety assurance of systems and products as ML becomes more widely used and interpretability of decisions made by complex systems becomes necessary (Jia et al 2021a). But we also see a broader role for counterfactuals which we consider as part of our proposed re-framing of system safety engineering.

It is often stated that safety engineering is a through-life concern but, in practice, much of the emphasis is on pre-deployment analysis of a particular system – with the assumption that systems will remain safe (enough) through life if well-maintained and operated appropriately. It is suggested here that the through-life nature of safety engineering needs to be widened and reinforced to consider very early life-cycle issues (concept design), and through-life system monitoring, e.g.

the lifetime environmental impact from a vehicle, including disposal, as well as accident and incident analysis.

First, is to incorporate "counterfactual thinking" at the earliest stages of system design, considering major design options. For example, if road infrastructure is modified to include inductive charging, then electric vehicle battery sizes could be reduced. There is an environmental impact (harm) in reworking existing roads and the implications for embodied carbon[7] to set against the benefits of reduced battery requirements, and the consequent change to the supply chain (including factories). By considering "what is not but could have been" for the major options it will give a basis for reasoning about the "best" alternative.

Practical trade-offs tend to compare possible changes to a baseline design – a similar approach could be adopted here, looking at the "delta" in benefits and harms in comparing the proposed design with alternatives (this is consistent with the way counterfactuals are generated in ML, seeking to minimise the change in inputs to produce the desired outputs). These trade-offs would need take a through life perspective, and it may be that approaches from economics or healthcare, such as Quality Adjusted Life Years (QALYs)[8], would be useful. Given the breadth of considerations in this re-framing of safety engineering, there will be many stakeholders – and the majority of these will be "lay", in the sense of not being specialist in the technology. Again, we would see the counterfactuals as having a role – "we don't recommend this option because …". However, the environment can't "speak for itself" and there will be a need to seek out appropriate stakeholders covering all the relevant concerns.

Second, it is important to monitor systems in operation to see whether or not they behave as predicted – including in safety terms. A number of projects are exploring the notion of learning from operations, including through use of digital twins, for example at the Alan Turing Institute[9]. It is possible to apply ML to operational data to identify cases where system behaviour deviates from what was predicted in a way that has implications for safety (Jia et al 2021b). Further, with very complex systems it is hard to predict all the possible behaviours of the system in advance and its wider impacts in terms of society and the environment (McDermid et al 2021). Here the need is to monitor for continuous changes/long-term trends, not events. We suspect that counterfactual approaches could help here – for example identifying the minimum set of changes necessary in the system or its operation to achieve the intended balance between benefits and harms.

Third, effective learning from accidents and incidents includes a form of counterfactual reasoning. Section 2.1 illustrated the long-term safety improvements that have been achieved in air transport. One of the reasons for this success is the

---

[7] See: https://www.raeng.org.uk/RAE/media/General/Policy/Net%20Zero/NEPC-Policy-Report_Decarbonising-Construction_building-a-new-net-zero-industry_20210923.pdf

[8] See: https://www.nice.org.uk/glossary?letter=q

[9] See: https://www.turing.ac.uk/research/research-projects/theoretical-foundations-engineering-digital-twins

thoroughness of accident investigation in seeking the underlying root causes of accidents, not just the proximate cause. The above-mentioned emphasis on CRM (crew/ cockpit resource management) is a case in point. Although not usually expressed this way, learning from experience in safety is concerned with finding actionable counterfactuals – if x had not occurred, y would not have occurred; therefore, we need to remove the possibility of x or reduce its probability to avoid occurrence of an accident with a particular signature.

In addition, many modern systems are data rich. It may therefore be possible to see the chain of events that led up to an accident in the data and to use explainable AI (XAI) methods to generate counterfactuals during analysis of accidents and incidents – identifying the minimum changes that could have prevented the accident. Care is needed, however, in that ML identifies correlations in data, not causation, so all learnt models and suggestions need to be subject to human scrutiny. For example, a counterfactual that says an aircraft would have avoided a runway overrun if it hadn't landed is true but unhelpful. One interesting area of work is on contrastive explanations (Lipton 1990), including identifying "pertinent negatives": factors whose *absence* is necessary to draw a particular conclusion (Dhurandar et al 2018). A contrastive explanation that compares accident-free behaviours with accident scenarios might identify missing controls – pertinent negatives – that, if present, could have prevented the accident from arising. Note that this will not work in all cases. If the controls needed are new – not already an aspect of system design or operation – then this approach will not find them. Counterfactual thinking is still valuable, but automated ways of generating counterfactuals will not be a panacea.

Finally, whilst we have indicated a role for ML in supporting system monitoring and accident/incident analysis, we note that the data centres that support on-line services, including ML applications, account for about 1% of global electricity supply[10]. Whilst this is not all due to ML, this environmental impact needs to be considered and suggests, again, the need to think holistically. The reframed safety engineering framework, focused on a broad notion of 'well-being' – and with a central role for counterfactual explanations, analysis, and reasoning – ought to be a powerful tool in shaping policy and choosing amongst policy options, including allowing for their longer-term impact.

5 Discussion and Conclusions

In this paper we have striven to be bold yet realistic! We have highlighted some challenges for a modern conception of system safety, starting from the contradictions of consumerism. Increased availability of safe products and services, for

---

[10] See: https://www.iea.org/reports/data-centres-and-data-transmission-networks

example cars and air travel, leads to environmental damage — and ultimately deleterious effects on human safety, health, and well-being. In light of such concerns, we have suggested steps for a progressive re-framing of safety engineering. This adopts the wider goal of 'well-being' as part of a more holistic, human-centred approach. The shift in focus covers societal and environmental impacts, as well as the impact both on individuals' psychological and physical well-being. Although the scope is wide, we believe our discussion shows the merit in a simple overarching structure to frame the concerns.

While there is other work on re-framing safety engineering, we don't believe any have the necessary scope to deal with the broad issues identified here. Safety II (Hollnagel 2018) focuses on complex socio-technical systems and on understanding "what goes right" as well as considering failures (which he characterises as Safety I). Undoubtedly this is a useful mindset, but it doesn't address the wider societal and environmental issues identified here.

Our ideas can be seen as endorsing the inclusion of a wider ethical perspective in the evolution of safety engineering. There are several initiatives around ethics of autonomy, for example the IEEE's work on Ethically Aligned Design and their P7000 series of standards.[11] The P7000 initiative is important and visionary, and includes documents addressing well-being (IEEE 2020), but the work is restricted to autonomous and "intelligent" systems. As should be clear from the above examples, the scope of concern here is much broader. The approach we have outlined should give a basis for realising ethically aligned design on a broader front than just autonomous systems – although that remains to be demonstrated.

The re-framing of safety engineering that has been suggested in this paper is broad. If it can be realised at all, then it can only be achieved over time. Some of the more speculative ideas need to be assessed for feasibility and we see this paper as the "start of a journey" not a well-specified destination. There are also open questions for future work and systematic reflection which should guide that journey:

1. Can safety engineers address all these concerns (individual, societal, environmental impact) alone? If so, how do they achieve the necessary knowledge? Alternatively, should safety engineers take on a role of integrating thinking from a range of disciplines, perhaps with an emphasis on articulating the trade-offs between incommensurable concerns?
2. How should we approach defining the objective (at least calculable) measures of well-being and acceptable risk? Again, there is a need to draw on the theoretical resources of other disciplines, perhaps econometrics or healthcare (e.g. adopting or adapting the QALY) but can general risk classes be defined, or do risks need to be evaluated on a case-by-case – system-by-system – basis?

---

[11] For the list of P7000 standards can be found see: https://ethicsinaction.ieee.org/p7000/

3. How can we reason about distributions in risks (harms) and benefits from (new) systems, and how do we engage relevant stakeholders in decision-making? This is likely to need methods of stakeholder engagement from social science. Counterfactual explanations might also be useful in communicating about alternative possibilities to a diverse, and lay, audience.

In our view, the need for re-framing system safety is pressing. The ideas presented here are intended to stimulate debate and help to influence the future evolution of safety engineering. One thing is clear; this must be a multidisciplinary undertaking. Ethicists, psychologists, environmentalists, human factors specialists, and experts in supply chains all need to be involved, almost regardless of the system considered. It is also likely that data scientists and experts in AI/ML will make a substantial contribution to developing practicable new analysis methods, especially when considering operational data. In specific domains, e.g. aviation, other specialists, e.g. atmospheric chemists, will need to be involved. Our hope is that we have provided enough of a starting point to enable these disparate groups to start to work together within an enlarged, human-centred framework for safety engineering.

Acknowledgements This work was supported by the Assuring Autonomy International Programme, funded by the Lloyd's Register Foundation and the University of York.

References

Automobile Association (AA) (2021) From windscreen wipers to crash tests and pedestrian protection https://www.theaa.com/breakdown-cover/advice/evolution-of-car-safety-features. Accessed 11 October 2021

Avizienis A, Laprie J-C, Randell B and Landwehr C (2004) Basic concepts and taxonomy of dependable and secure computing, in IEEE Transactions on Dependable and Secure Computing 1(1):11-33,

Berners-Lee M and Clark D (2010) What's the carbon footprint of ... a new car? https://www.theguardian.com/environment/green-living-blog/2010/sep/23/carbon-footprint-new-car. Accessed 7 September 2021

Boeing (2020) Statistical summary of commercial jet airplane accidents: worldwide operations, 1959-2020. https://www.boeing.com/resources/boeingdotcom/company/about_bca/pdf/statsum.pdf. Accessed 11 October 2021

Burnett R et al (2018) Global estimates of mortality associated with long-term exposure to outdoor fine particulate matter. Proceedings of the National Academy of Sciences 115(38): 9592-9597

Chen Y et al (2021) A review of lithium-ion battery safety concerns: The issues, strategies, and testing standards. Journal of Energy Chemistry 59: 83-9

COMEP (2018) Associations of long-term average concentrations of nitrogen dioxide with mortality, Public Health England 2018238

Crisp R (2021) Well-Being, The Stanford Encyclopaedia of Philosophy (Fall 2021 Edition), Zalta E (ed.). https://plato.stanford.edu/archives/fall2021/entries/well-being. Accessed 11 October 2021

Department of Energy, U.S. (2014). The history of the electric car. https://www.energy.gov/articles/history-electric-car. Accessed 11 October 2021

Department for Transport, G.B. (2020) Reported road casualties in Great Britain: 2019 annual report https://assets.publishing.service.gov.uk/government/uploads/system/uploads/attachment_data/file/922717/reported-road-casualties-annual-report-2019.pdf. Accessed 11 October 2021

Doyal L and Gough I (2011) A theory of human needs. MacMillan

Dhurandhar, A., Chen, P.Y., Luss, R., Tu, C.C., Ting, P., Shanmugam, K. and Das, P. (2018) Explanations based on the missing: towards contrastive explanations with pertinent negatives. arXiv preprint arXiv:1802.07623.

Hassel H and Cedergren A (2021). A framework for evaluating societal safety interventions, Safety Science, 142:105393

Hollnagel E. (2018). Safety-I and Safety-II: the past and future of safety management. CRC press.

IEEE (2019) Ethically Aligned Design: A vision for prioritizing human well-being with autonomous and intelligent systems. The IEEE Global Initiative on Ethics of Autonomous and Intelligent Systems. https://standards.ieee.org/content/ieee-standards/en/industry-connections/ec/autonomous-systems.html. Accessed 11 October 2021

IEEE (2020) IEEE P7010-2020, IEEE recommended practice for assessing the impact of autonomous and intelligent systems on human well-being. https://standards.ieee.org/standard/7010-2020.html. Accessed 11 October 2011

IPCC (2021) Climate change – widespread, rapid and intensifying, https://www.ipcc.ch/2021/08/09/ar6-wg1-20210809-pr/. Accessed 6 September 2021

Jia Y, McDermid, J A, Lawton T and Habli I (2021a) The role of explainability in assuring safety of machine learning in healthcare. Submitted to IEEE Transactions on Emerging Topics in Computing (available at: arXiv preprint arXiv:2109.00520).

Jia,Y, Lawton T, McDermid J A, Rojas E and Habli I (2021b) A framework for assurance of medication safety using machine learning. arXiv preprint arXiv:2101.05620.

Kang D, Chen M, Ogunseitan O (2013) Potential environmental and human health impacts of rechargeable lithium batteries in electronic waste. Environ Sci Technol 47(10):5495-5503.

Kulp S and Strauss B (2019). New elevation data triple estimates of global vulnerability to sea-level rise and coastal flooding. Nature communications 10(1): 1-12.

Lipton P (1990) Contrastive explanation. Royal Institute of Philosophy Supplements 27:247-266.

Lyth P J, Dierikx M L J (1994) From privilege to popularity: the growth of leisure air travel since 1945. J Transport History 15(2):97-116

McDermid JA, Burton S, Garnett P, Weaver RA (2021) An initial framework for assessing the safety of complex systems, in Parsons M and Nicholson M (eds). Systems and COVID-19: Proceedings of the 29th Safety-Critical Systems Symposium Virtual Conference,

Miller T (2019) Explanation in artificial intelligence: insights from the social sciences. Artif. Intell. 267: 1–38

Nussbaum M and Sen A (1993) Capability and Well-being, in Nussbaum M and Sen A (eds.) The quality of life. Clarendon Press

Owen R, Bessant J. Heintz, M. eds. (2013) Responsible innovation: managing the responsible emergence of science and innovation in society. John Wiley & Sons.

Parfit D (1984) Reasons and persons. Oxford University Press

Ritchie H (2021) Cars, planes, trains: where do CO2 emissions from transport come from? https://ourworldindata.org/co2-emissions-from-transport. Accessed 6 September 2021

Schumaker E F (1966) Buddhist economics in Asia: a handbook, Wint G (ed., Anthony Blond Ltd. https://web.archive.org/web/20121213145110/http://neweconomicsinstitute.org/buddhist-economics Accessed 8 September 2021

Starr W (2021), Counterfactuals, The Stanford Encyclopedia of Philosophy (Summer 2021 Edition), Edward N. Zalta (ed.) https://plato.stanford.edu/archives/sum2021/entries/counterfactuals. Accessed 11 October 2021

Statista (2021) Number of scheduled passengers boarded by the global airline industry from 2004 to 2020 https://www.statista.com/statistics/564717/airline-industry-passenger-traffic-globally/. Accessed 6 September 2021

The Sunday Times (2020) Vehicle emissions in the spotlight again as coroner concludes air pollution contributed to death of schoolgirl https://www.driving.co.uk/news/environment/air-pollution-contributed-death-nine-year-old-coroner-rules/. Accessed 7 September 2021

World Bank (2021) Air transport, registered carrier departures worldwide https://data.worldbank.org/indicator/IS.AIR.DPRT. Accessed 11 October 2021

VividEconomics (2018) Accelerating the EV Transition Part 1: environmental and economic impacts. https://www.wwf.org.uk/sites/default/files/2018-03/Final%20-%20WWF%20-%20accelerating%20the%20EV%20transition%20-%20part%201.pdf. Accessed 11 October 2021

Wachter S, Mittelstadt B, Russell C (2018) Counterfactual explanations without opening the black box: automated decisions and the GDPR. Harvard Journal of Law & Technology 31(2).

Von Schomberg R (2013) A vision of responsible research and innovation, in Owen R, Bessant J, and Heintz M,(eds) Responsible innovation: managing the responsible emergence of science and innovation in society. John Wiley & Sons

Does an agile approach improve the way we derive RAMS requirements?

Gavin Wilsher

Capgemini Engineering

Abstract *Agile methods appear to offer a more dynamic way to drive a project forwards. So, can we derive RAMS requirements in a more efficient manner when working within an agile systems development environment? Recent experience has provided some interesting insights into whether improvements can be realised and how we might capitalise on the more iterative nature of an agile systems development environment. This paper explores recent experiences in the early stages of rail development projects (EN50126 phases 3-5). Applying both a Scrum methodology and a Kanban approach provided two slightly different perspectives from which to learn. This paper looks at the benefits and pitfalls of deriving RAMS requirements within an agile systems development framework. Lessons have been learned around systems modelling, use case development and specialist discipline integration and this paper presents a series of recommendations based on those lessons.*

1 Introduction

Agile promises much in terms of efficiency in the development of software, but can we use it in a systems development environment to help us to derive Reliability, Availability, Maintainability and Safety (RAMS) requirements in a more efficient and effective way than using more traditional approaches?

This paper sets out to explore the author's recent experiences of working in an agile development environment and hopes to provide at least one answer to the question that might be useful to other project teams embarking on this journey. This is not intended to be a learning-from-experience paper for a single project since that would probably not be of benefit, or interest, to the reader. However, it does represent the author's views based on a variety of safety-related project experiences and seeks to offer some (hopefully) useful guidance on how one might ensure that maximum benefit is gained when a project decides to follow an agile approach.

© Gavin Wilsher 2022.
Published by the Safety-Critical Systems Club. All Rights Reserved.

2 Context

Agile is more usually associated with software development. It began as a software development methodology aimed at the rapid development of software applications and web applications, situations where there is no hardware and certainly no safety analysis required. However, it is becoming widely adopted in the systems engineering environment and is also being applied in safety-related and safety-critical projects. A useful text with regard to its application in systems engineering environments is *Agile Systems Engineering* (Bruce Powell Douglas, 2015).

In the context of the author's experience and for the purposes of this paper, the subject projects are mainly systems developments (incorporating software) rather than software development alone and all are either safety-related or safety-critical systems. Of course, in the early stages of system development, some software requirements will be derived but they are not the main focus of this paper.

It is worth noting that the particular flavour of agile that was adopted on these projects centred around behavioural use cases. This is explained further here but it should be noted that the author acknowledges that this is only one of many approaches that could have been taken under an agile framework.

The author had a number of questions going into these projects. These questions were partly born out of a lack of previous experience of agile, and partly due to a healthy scepticism of agile in safety-related / safety-critical developments based on reading some of Professor Nancy Leveson's work[1]. The following list presents these questions:

- RAMS requirements: is agile a better way to develop requirements?
- Quality assurance (QA): how do we ensure that robust QA is applied to the agile development?
- Safety: how do we derive hazards and safety requirements without taking a system level view?
- Security: do security requirements need to be based on the use cases?
- Traceability: how do we ensure end-to-end traceability?
- Verification and Validation (V&V): how can we do V&V iteratively without taking a final view at the "end" to ensure that V&V is complete?
- Documentation: agile likes to be light on documentation. How can that work in a regulated environment?
- Corner cases: does the agile approach increase or decrease the likelihood of missing a "corner case"?

---

[1] The most notable being *An Engineering Perspective on Avoiding Inadvertent Nuclear War* (Leveson, date unknown). For those who are not aware, Nancy Leveson is Professor of Aeronautics and Astronautics at MIT.

- Team size: agile claims to reach a solution to a client's problem more quickly. Does that mean we need a larger safety team, and does the same hold for the other specialists?

Some of these concerns turned out to be unfounded and others turned out to be justified. This paper intentionally only covers some of these points and specifically looks at the question that drove the author to write this paper. Something that has become abundantly clear over the course of these projects is that agile development methods in their pure form are not the "holy grail" that some of its more ardent supporters would have us believe. This is certainly true in the context of safety-related and safety-critical systems engineering projects.

3 Use Cases

As use cases are mentioned a few times in the paper it is worth providing a brief explanation of them.

The first point to note is that use cases are not related to agile. They are a standard systems engineering modelling tool used for capturing and modelling the behaviours and requirements of a system. They can be represented as UML[2] (or SysML[3]) diagrams, but can equally be textual in presentation. They are intended to describe the behaviours that the system under consideration needs to perform in order to contribute to its goals. Much of the author's understanding of systems engineering, and use cases in particular, has been learned from the academic paper *An Introduction to Systems Engineering with Use Cases* (Alexander I, Zinc T (2002) and the Systems Engineering Body of Knowledge (2021).

The basic use case diagram in Figure 2 shows a behavioural use case for a vehicle Anti-lock Braking System (ABS). The system has the driver, wheels and control module as actors[4] in the system and the requirements are traced to the use case. Once the functional requirements have been derived it is then possible to conduct a functional safety (and RAM) analysis of the use case.

---

[2] Unified Modelling Language

[3] Systems Modelling Language

[4] Actors are entities that interact with the system. They can be people, other systems or organisations.

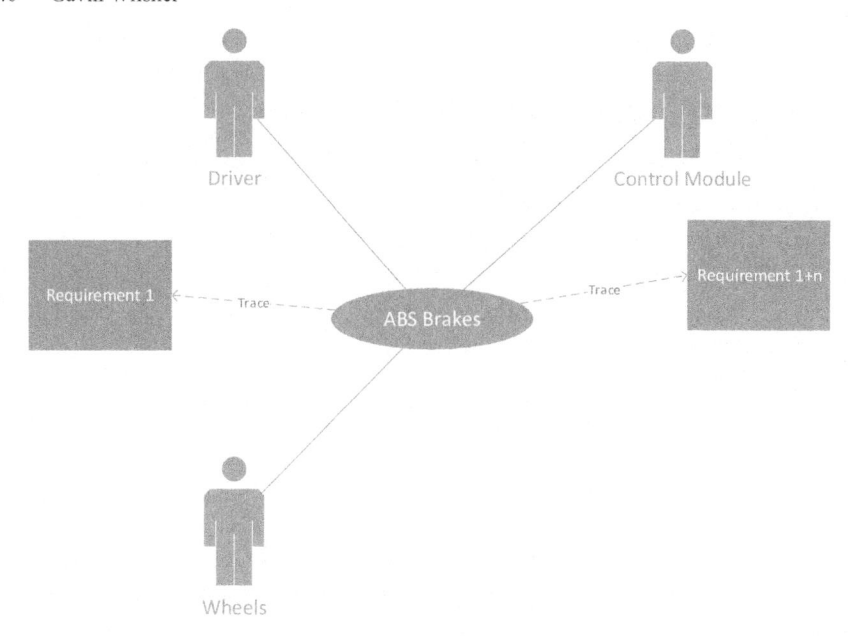

Fig. 1. Use Case Diagram

Key to diagram
- The green icons are the actors
- The dashed lines show the trace from the system to the requirements
- The blue boxes represent the requirements
- The blue oval represents the system at the heart of the use case

4 Scrum and Kanban – an understanding

This paper is certainly not intended to be a tutorial on agile, scrum, Kanban or systems engineering. However, some basic explanations are necessary in order to convey the message of this paper in the correct context.

Scrum and Kanban are slightly different approaches to agile development. The following sub-sections briefly describe what they are and, more importantly, how the author has seen them applied to a number of projects.

4.1 Kanban

Literally translated from Japanese, it means signboard or billboard. This term is used to describe the Kanban board that is used to track the workflow. The simplified diagram in Figure 2 illustrates a simple Kanban board. This is a method to manage and improve workflows.

It originated in lean manufacturing and was inspired by the Toyota 'Just in Time' production system (Wikipedia). The basic approach is to pull work through once the capacity to do the work is available, rather than pushing work forwards when there is no capacity to deal with it.

Despite being quite commonly used in software development projects today, the philosophy dates back to the 1940s and, as such, is nothing new!

In the context that the author has experienced it being applied, the Kanban approach ensured that each use case was fully developed before the team progressed to the next use case, thus ensuring all of the relevant subject matter experts were focused on getting the use case completed, or to use an agile term – getting it to "done". However, it could be considered that this is just a scaled down waterfall approach rather than anything beneficially new or improved.

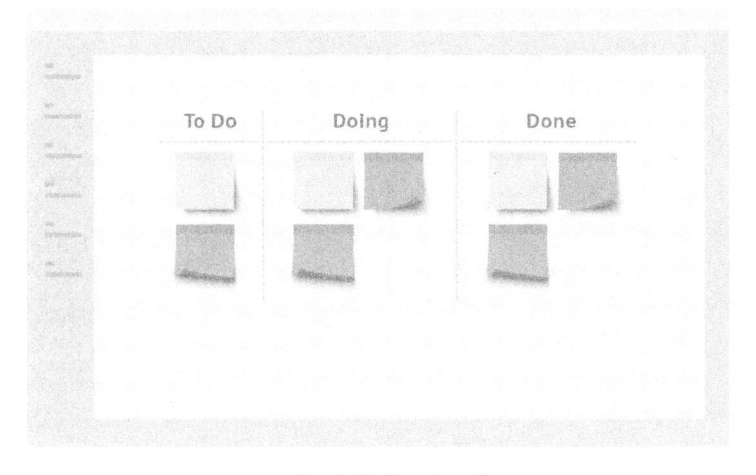

Fig. 2. Kanban board

4.2 Scrum

In the context the author has experienced it being applied, scrum provided a framework whereby the use cases could be developed in "sprints" i.e. time-boxed iterations, to derive system and sub-system level requirements simultaneously.

Scrum requires an 'agile mindset' and is centred on small self-organising teams breaking their work into chunks that can be completed within time-boxed iterations, called sprints, which are typically a couple of weeks long.

The scrum team assess their progress in time-boxed daily meetings called daily scrums. At the end of each sprint, the team holds two further meetings: the 'sprint review' which demonstrates the work done to stakeholders to elicit feedback, and 'sprint retrospective' which enables the team to reflect and improve.

The generic scrum diagram in Figure 3, copied from the Jose Lara "Medium" website[5] (Jose Lara, 2018), neatly illustrates the scrum approach. The product backlog is the work not yet started; the sprint backlog is the work that the scrum team has decided to take into the sprint i.e. the work that it (the scrum team) believes can be completed within the sprint; work is done during the sprint with the intent that you have something deliverable at the end of the sprint i.e. the finished work in Figure 3. The 'scrum master' oversees the scrum process.

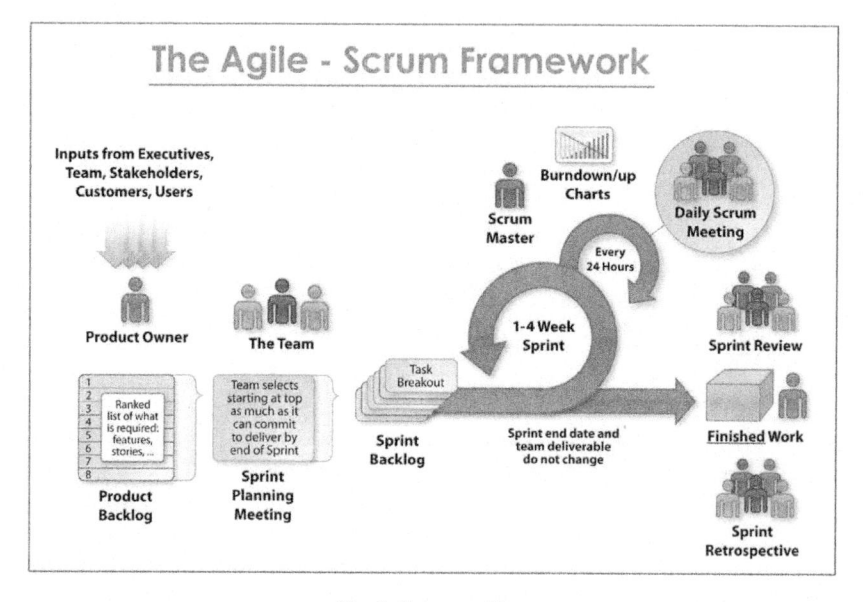

Fig. 3. Scrum outline

---

[5] No copyright is asserted by Jose Lara on this website.

5 What are we trying to deliver?

In the context of the projects that have informed this paper, EN50126[6] phases 3-5 are relevant. The *main* deliverables from these phases are:

- Phase 3:
  - Identify the safety hazards and the RAM equivalents;
  - Assess the risks;
  - Select risk acceptance principles.
- Phase 4:
  - Specify the system level RAMS requirements;
  - Specify the acceptance criteria for RAMS.
- Phase 5:
  - Apportion the system RAMS requirements to the subsystems;
  - Design the subsystems such that they work together as a system to fulfil the required system level functions;
  - Specify the interfaces for all subsystems derived from the RAMS requirements.

6 Can we deliver this iteratively? If we can, is it useful to the client?

In the context of these systems development projects, the three EN50126 lifecycle phases described above were delivered as a single project phase i.e. concurrently. This meant that system and subsystem requirements were being derived at the same time during each use case.

As discussed previously, use cases were developed and these were used to break the work up, with each being considered as a stand-alone package of work used to derive the requirements. Each iteration was a single use case. So in the broadest sense, it can be seen that it is possible to deliver requirements iteratively in an agile manner (or perhaps incrementally might be more accurate) once each use case has been completed.

However, it may be considered that this would have been of little value to the client since they would need the full requirements set to begin a procurement process and issue tenders for the system (and its sub-systems). The main benefit could be argued to be that it provided the client with visibility of the evolving requirements set and enabled them to comment on these in a timely manner rather

---

[6] EN50126: Railway Applications - The Specification and Demonstration of Reliability, Availability, Maintainability and Safety (RAMS): 2017

than have to wait until a full delivery. However, the iterative nature of the development could also have meant that subsequent use cases may have changed the requirements already delivered and reviewed because use cases can overlap each other.

7 Did it offer an improvement over a traditional waterfall approach?

Obviously, there are positive and negative sides to both approaches and it could be argued that the best approach in any system development is neither one nor the other but perhaps some combination of the two. The author has seen the combination of waterfall and agile approaches referred to as "Wagile" but is not suggesting a formal hybrid in that sense. Decades of experience has shown the author that the success (or failure) of a project can seldom be attributed to the overall approach adopted, but it (success or failure) can lie in the way that the chosen approach is implemented, and the constraints placed upon it.

The traditional waterfall approach, following the EN50126 standard lifecycle, facilitates a system level view from the outset whereas the scrum and Kanban agile approaches adopted, facilitated a much more detailed look at how different elements of the system interact with each other without necessarily considering the system as a whole.

8 Why is the system level view important?

When conducting safety analysis to derive safety requirements, it is vital that the boundary of the system is considered (after all that is where the hazards reside). Figure 4 illustrates the concept of hazards being at the boundary of the system under consideration, with causes to left of the system boundary and hazards at the railway system level and accidents to the right.

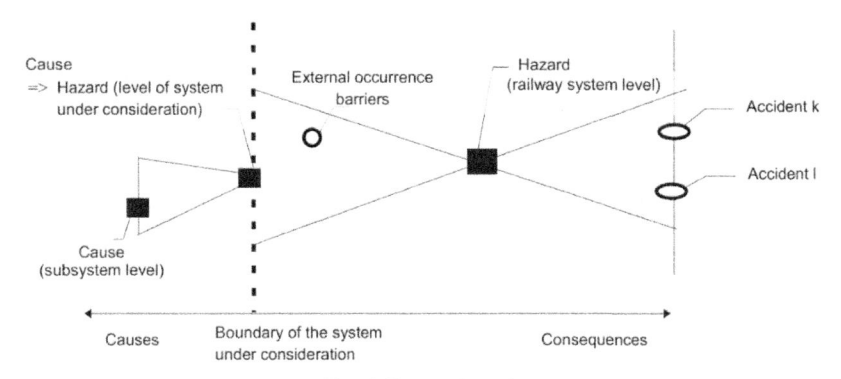

Fig. 4. System Boundary

In taking a use case view without the benefit of being able to consider the system boundary, it is not always possible to derive safety requirements that will mitigate system level hazards at the system boundary (because these are unknown). A similar conundrum arises when deriving RAM requirements without an understanding of how a functional failure might impact the whole system rather than just the subsystems involved in the use case. This makes it very difficult to derive system level RAM requirements to control the effects of failures at the system level. Overall, this means that it makes it challenging to make decisions that are best for the overall system without having full visibility of what the overall system actually is.

As an example of this; if one considers the ABS subsystem use case example in Figure 1, and further considers that the system boundary is the vehicle to which it is fitted, it can be seen that this use case touches what could be considered to be the boundary of the system (the vehicle's wheels). Consequences of failures in the ABS subsystem (or failures on the part of the actors) will be seen at the wheels and it is therefore possible to derive hazards at the system boundary and produce system level safety requirements to mitigate them. These can then be apportioned, as necessary, to the relevant subsystems which would of course include the ABS subsystem. This fits in neatly with the EN50126 approach in phases 4 & 5.

However, not all use cases will touch the system boundary in this way. The problem then becomes one of having an understanding of what the effects of a functional failure are at the boundary of the use case rather than the boundary of the system. This can easily result in producing subsystem requirements that have not been apportioned from system requirements i.e. orphans that may not serve to mitigate system level hazards. This does not fit in neatly with the EN50126 approaches in phases 4 & 5 because it addresses phase 5 without first addressing phase 4. Furthermore, this runs the risk of completing phase 5 without ever being certain whether completing phase 4 first would have resulted in more (or different) subsystem requirements being apportioned. This potentially presents a serious validation problem, which brings us back to the V&V question that the author had at the outset – can validation really be done iteratively without taking a final view at the end of the project to ensure that it is complete. We need to be sure that requirements derived in one use case have not been broken by requirements developed in subsequent use cases.

This is the point at which it *could* be argued that taking a use case *only* approach (within an agile systems development) definitely does *not* improve the way we derive RAMS requirements. However, there is always a way to improve things, especially with the benefit of hindsight, and the purpose of this paper is not just to highlight some of the issues that the author has experienced but to present some potential solutions to them that others might benefit from.

9 What could have been more useful?

Based on the experience gained from working on these projects, it can be seen that there were things that could (or should) have been done differently whilst still benefitting from the agile systems development approach adopted, certainly from a RAMS requirements derivation perspective.

The following paragraphs explore some of the potential improvements.

Functions. Agreeing on the high-level functions is an important aspect of performing functional failure analysis which can be overlooked when forging ahead with use case development. This is a high-level (system-level) activity that would be beneficial to complete prior to beginning iterative agile development.

System boundary. Amongst other things, EN50126 phase 2 requires that a project defines the system, its boundary, operational profile and establishes the operational requirements influencing the characteristics of the system. This useful exercise should form the starting point for RAMS analysis i.e. working at the system boundary and using the system definition and operational profile to identify safety hazards (and the RAM equivalent), assess risks and derive system level RAMS requirements i.e. following phases 3 and 4 as per the EN50126 standard. This then naturally flows into phase 5 – apportioning the system level requirements to the subsystems. Attempting to derive RAMS requirements without starting at the system level is not only contrary to the spirit of the standard, it simply makes things more difficult than they need to be.

Use cases. The use case work provides an opportunity to further refine the requirements, based on the specific behaviours modelled in each of the use cases. This is particularly useful for deriving further subsystem RAMS requirements that may not have been clear when considering the system level operational profile. At this point, it seems clear that iterating the process around different use case works as well for RAMS as it does for the derivation of non-RAMS functional system requirements. Whether delivering these newly derived requirements iteratively is of value to the client is again questionable (recall that one of the claimed benefits of agile is to deliver something useful to the client at regular intervals).

Modelling. When following an agile Model Based Systems Engineering (MBSE) approach, there may be a justifiable case for not waiting for the model to be complete and performing safety analysis on the system and its functions and then using the model to validate the analysis later on. If the model is evolving as part of the agile development then this would be beneficial after using the use cases to refine the requirements.

10 Final thoughts

This paper has highlighted some of problems that the author has experienced whilst working on the railway signalling system development projects briefly described in the context section of this paper and suggests some ideas that might improve things for future projects.

The author remains sceptical about the usefulness of pure agile in a systems development environment where the system is safety-related / safety-critical, especially in the early phases of the lifecycle.

There were some notable benefits to the approach taken, although they don't necessarily serve to make the derivation of RAMS requirements more efficient or effective. However, they are worth highlighting:

- The RAMS team is fully integrated into the systems engineering / development team during the use case development. This is not always the case and ensures that RAMS (most notably safety) is able to lead in the design effort, resulting in safety being more integrated into the system design rather than being considered an add-on, as is often the case on projects;
- Specialist stakeholders from the client organisation can be embedded in the team;
- RAMS analysis can be done as a team activity without the need to convene workshops because the entire team, including the client specialists, are working together continuously.

With the implementation of the suggestions in section 9 of this paper, there is certainly good reason to believe that applying iterative use cases in an agile environment to EN50126 phase 5 could offer an improvement over the tried and tested approach to deriving RAMS requirements. However, that would depend upon only adopting agile once the project is well underway i.e. starting the project with a more traditional approach, certainly up to EN50126 phase 5, and only then introducing agile. If the chosen approach is one of applying agile to use cases, then these would serve to refine the RAMS requirements already developed following a more traditional approach.

Perhaps then, the author does not fully concur with Nancy Leveson's rather controversial view that agile does not belong or serve a useful purpose in the world of safety-related / safety-critical systems development. Although her view would certainly seem to be true if agile is applied from the outset and is not modified in some way to take into account the need of RAMS to take a whole system view in the early phases of a project.

So to re-visit the question posed by this paper *"Can working in an agile systems development environment improve the way we derive RAMS requirements?"* the answer is yes, but only if the application of agile is adapted to satisfy the need of RAMS to take a whole system view from the outset.

The independent research organisation, SINTEF has developed an approach called *SafeScrum* (Stålhane T, 2013). The basic concept of the approach is that it splits the software development from the rest of the development process so that only the software development is handled by the *SafeScrum* process, with the rest e.g. systems design, safety decisions and new safety requirements, being kept outside scrum. The *SafeScrum* process also introduces a second backlog to the standard scrum process – the safety product backlog, acknowledging the fact that the safety products need to be handled differently to the functional products.

The basic concept of *SafeScrum* seems rather similar to the approach being suggested in this paper i.e. that some aspects of development do not lend themselves readily to the agile approach and are probably best kept outside it. Although this paper suggests that the separation may only be necessary in the early phases of development.

One final thought – perhaps the RAMS standards currently followed should be updated to consider how they might be applied effectively in an agile environment or perhaps they should at least take a stance on the applicability of agile in safety-related / safety-critical systems development.

Acknowledgments The author would like to acknowledge Stu Tushingham (Capgemini Engineering) for his guidance and review input. The author would also like to thank Andrew Hawthorn (Capgemini Engineering) for his review input.

Disclaimer This paper represents the author's thoughts, opinions and views and does not represent the formal position of Capgemini Engineering in relation to agile.

References

1. https://medium.com/@realjoselara/agile-scrum-process-in-a-nutshell-6ec32a59efb (accessed 26/10/2021)

2. Guide to the Systems Engineering Body of Knowledge, SEBoK v. 2.5, released 15 October 2021
3. Alexander I, Zinc T (2002) An Introduction to Systems Engineering with Use Cases
4. Professor Nancy Leveson An Engineering Perspective on Avoiding Inadvertent Nuclear War
5. Bruce Powell Douglass (2015) Agile Systems Engineering
6. https://en.wikipedia.org/wiki/Kanban_(development) (accessed 9/11/2021)
7. Stålhane T (2013) Safety standards and Scrum – A synopsis of three standards, https://www.sintef.no/globalassets/safety-standards-and-scrum_may2013.pdf accessed January 2022

Design-time Specification of Dynamic Modular Safety Cases in Support of Run-Time Safety Assessment

Elham Mirzaei[1], Carmen Cârlan[2], Carsten Thomas[1], Barbara Gallina[3]

[1] HTW Berlin, University of Applied Sciences, Berlin, Germany

[2] fortiss GmbH, Munich, Germany

[3] Mälardalen University, Västerås, Sweden

Abstract *Open Adaptive Complex Systems – such as road vehicle platoons or fleets of cooperative robots – may use dynamic reconfiguration to adapt to system or environment changes. One approach enabling this feature is Service-oriented Reconfiguration, where new configurations are created by composing the available services in an unconstrained manner. Due to the high number of possible service compositions, not all configurations can be pre-assured at design-time. Despite recent progress, there is no satisfactory approach for specifying safety cases in support of their re-evaluation at run-time, after system reconfiguration. To this end, in previous work, we introduced Dynamic Modular Safety Cases (DMSC). A DMSC is a modular safety case, which can be dynamically re-constructed and re-assessed given service reconfiguration. In continuation of the previous work, in this paper we provide guidelines for specifying safety cases at design-time, whose modular structure mirrors the system service decomposition, to enable their re-construction and re-evaluation at run-time in the event of a system reconfiguration. Aiming to support the specification of DMSC, we extend FASTEN, an engineering tool for the design and verification of safety-critical systems. We exemplify the specification of DMSCs in FASTEN for an illustrative example from the smart factory domain.*

1 Introduction

System safety, i.e., the fact that the system deployment does not pose an unacceptable risk of harm, needs to be assured. Assurance here is interpreted as *"grounds for justified confidence that a claim has been or will be achieved"* (ISO/IEC/IEEE 2019). In most safety-critical domains, such as automotive, or

© Elham Mirzaei, Carmen Cârlan, Carsten Thomas, Barbara Gallina, 2022.
Published by the Safety-Critical Systems Club. All Rights Reserved.

nuclear, or rail system safety assurance also assumes the creation of a system safety case. A Safety Case (SC) is defined as *"a reasoned and compelling argument, supported by a body of evidence, that a system, service or organisation will operate as intended for a defined application in a defined environment"* (Ministry of Defence 2007). Despite some criticism against SCs (Leveson 2020), they are widely applied in various domains, and their provision may be even required (e.g., air traffic control (Eurocontrol 2006), road and rail transportation (CENELEC 2007)). Typically, SCs are created at design-time and maintained throughout the life of the system.

Open adaptive complex systems (often encompassing Systems of Systems), such as road vehicle platoons within the automotive domain and fleets of cooperative robots within the robotics domain, are characterized by dynamic evolution/reconfiguration and emergent behaviour (Boardman and Sauser 2006). Very often, these systems are safety-critical. Hence, maintaining the safety assurance of such system after dynamic changes that are imposed by reconfiguration is necessary (Kelly 2003).

In the literature, reconfiguration is often classified into three types, namely *predefined*, *constrained*, and *unconstrained selection* (Bradbury, et al. 2004). Unconstrained selection provides more flexibility in terms of adaptation at run-time amongst all possible variations for creating a new configuration. Open adaptive complex systems use reconfiguration to adapt their structure and behaviour to changes in constituent systems or in their environment. In earlier work, we have introduced the Service-oriented Reconfiguration (SoR) approach (Wudka, et al. 2020) (Thomas, et al. 2021), which supports unconstrained reconfiguration at run-time based on the Service-oriented Architecture (SoA) concept and blueprints.

The SCSC[1] Service Assurance Working Group (SAWG) provides guidance on challenges related to the safety assurance of services, e.g., inter-service interference, mapping from service decomposition to modules within modular assurance, and deviation to the reference architecture due to the change in configuration during deployment (SAWG 2020). They highlight the necessity of extending beyond the traditional approaches in system safety engineering to address those challenges, considering that the future developments in business and technology are likely to adopt this service paradigm in the next generation of safety-critical complex systems.

In the context of reconfiguration classifications, unconstrained reconfiguration is the most challenging approach from the safety assurance perspective. This is mainly because, in current practice, a SC is developed manually, by a safety engineer at design-time, compiling evidence produced during the execution of safety assurance activities. As the argumentation structure comprised by the SC largely depends on the system configuration, the current manual approach for SC

---

[1] Safety-Critical Systems Club

development at design-time is inappropriate for SoR-based systems, for which knowing all possible configurations prior to operation is difficult. There is a need to enable automated development and assessment of the system safety case at run-time, considering the run-time system reconfiguration.

To this end, in our previous work, we introduced the DMSC approach (Mirzaei, Thomas and Conrad 2020), which combines two state-of-the-art approaches for developing safety cases: Modular Safety Case (MSC) and Dynamic Safety Case (DSC). MSC address challenges such as system complexity (Kelly 2003) and frequent system evolution by breaking down a safety case into several connected modules, and DSC aim at re-evaluating the validity of design-time safety assumptions at run-time (Denney, Pai and Habli 2015). The DMSC approach unites the modular structure concept and the concept of dynamic update and re-evaluation at run-time.

In this paper, we provide guidelines for constructing DMSC. The core idea behind these guidelines is to construct the SC at design-time in a manner that enables automatic SC re-construction and re-evaluation at run-time. Further, we also show how FASTEN – an engineering platform for the creation and maintenance of safety cases – has been extended to support the specification of DMSC and the management of their relations with the structural elements of open adaptive complex systems. Finally, we show how to develop DMSC following our proposed guidelines and using the FASTEN tool for an illustrative example from the smart factory domain.

The rest of this paper is organized as follows. In Section 2, we discuss the language we use in this paper for the specification of SC, and we briefly describe the reconfiguration approach we consider, highlighting the requirements it imposes with respect to the system safety assurance. Section 3 provides a brief overview of available methods for the creation and management of the SC. Section 4 outlines our proposed solution for the specification of DMSC. In Section 5, we present tool-support for the proposed solution, and we exemplify its usage for a small use case involving two robots in a factory. Finally, in Section 6, we summarize our contributions and outline the next steps.

2 Fundamentals

2.1 The GSN/SACM Metamodel

To better structure the SC, graphical notations have emerged over the past years, one of the most frequently used notations being the Goal Structuring Notation (GSN) (ACWG-GSN 2021). In GSN, structured SC arguments are constructed by goals, strategies, solutions, assumptions, and context definitions. Usually, a top-level safety goal is first defined, which is later decomposed to sub-goals and the step of goal decomposition is re-iterated until the evidence for the satisfaction of the sub-goals is referenced.

GSN supports the specification of MSC (Industrial Avionics Working Group 2012), a concept on which our DMSC approach relies, by introducing away goals, which are repeating a claim presented in other modules, thereby creating a reference from one argument to another. Fig.1 presents an MSC using GSN elements.

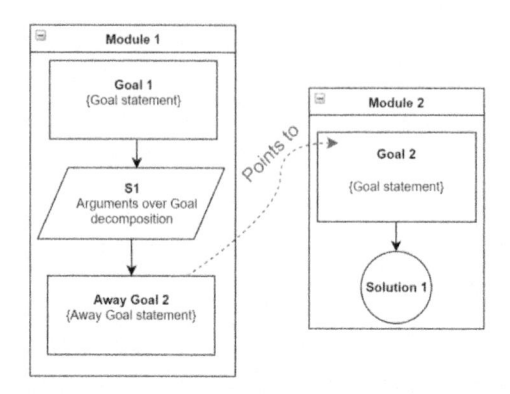

Fig. 1. A Modular Safety Case modelled in GSN using an Away Goal

MSC have been proposed to address challenges such as system complexity (Kelly 2003) and frequent system evolution by breaking down a safety case into several connected modules. If the safety case modularization is done appropriately, the modular structure of the safety case limits the change impact propagation only to a certain part of the safety case. Consequently, to update the safety case in the event of system changes, it may be only necessary to change certain modules, rather than the entire SC. This leads to the reduction in the cost and efforts on SC changes (Kelly and Bates 2005).

To improve standardization and interoperability between tools which support the modelling of GSN-based safety cases, the GSN/SACM metamodel was introduced (ACWG-GSN-MM 2021). GSN/SACM maps the GSN constructs to the SC elements described by the Structured Assurance Case Metamodel (SACM) (Object Management Group 2020) proposed by the Object Management Group (OMG). Among others, the GSN/SACM metamodel allows the specification of traces from safety case elements to other artefacts. As an example for the relation between the GSN notation and SACM, see the mapping of a GSN SC fragment (Goal G1 in Fig.2) to its SACM representation (Fig.3), taken from (ACWG-GSN-EX 2021).

> **G1**
>
> System X can tolerate single component failures

Fig. 2. An example SC fragment using GSN notation (ACWG-GSN-EX 2021)

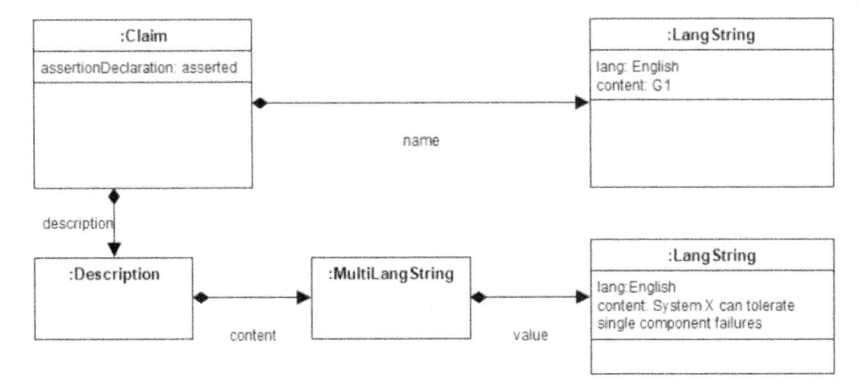

Fig. 3. SACM representation equivalent to the SC fragment shown in Fig. 2 (ACWG-GSN-EX 2021)

2.2 Service-oriented Reconfiguration (SoR)

In this subsection, we recall the basic elements of our Service-oriented Reconfiguration (SoR) approach to provide the context for the DMSC specifications, which will be introduced in Section 4.

Service-oriented Architecture (SoA) is a software design pattern that supports system modularization and interaction between system components (Richardson 2018). Applying this pattern to open adaptive complex systems (Siefke, et al. 2020), we perceive these to be composed of constituent systems (e.g., a fleet of robots consisting of individual autonomous robots), where the constituent systems realize their intended functions by sets of interconnected services for sensing, computation and actuation. Open complex adaptive systems may react to internal changes (e.g., malfunction behaviour) and changes in their environment by flexibly adapting the interconnections between services at run-time, within individual constituent systems or across several constituent systems. The SoR approach (Thomas, et al. 2021) builds on the SoA pattern to define a reconfiguration mechanism that – at run-time – creates new configurations for open adaptive systems using service blueprints to specify potential service network configurations.

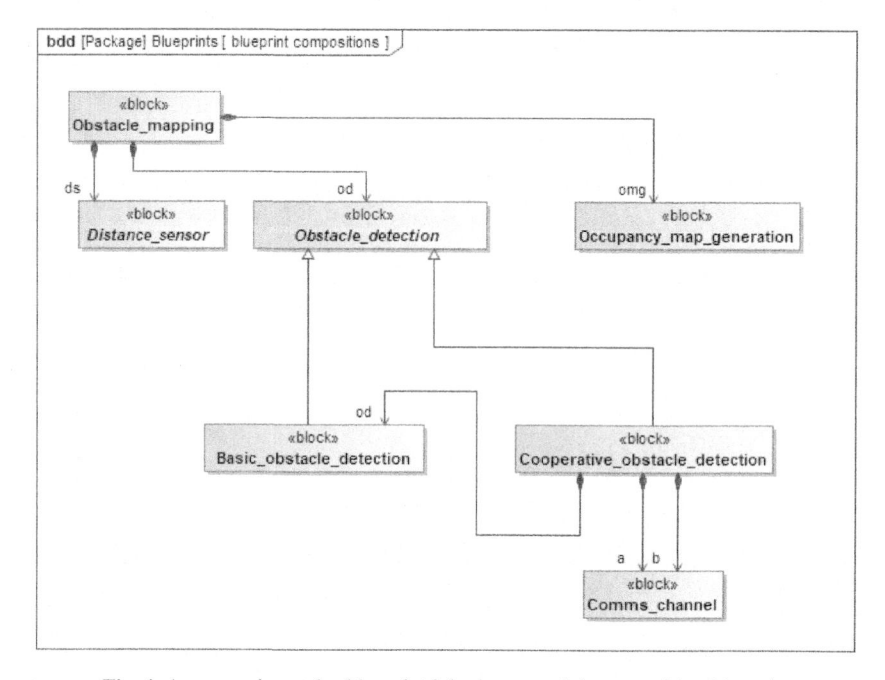

Fig. 4. An example service blueprint inheritance and decomposition hierarchy

A service blueprint is a template defining potential services, either as basic services or as service compositions. Depending on the type of blueprints, the specifications may include interface, parameters, internal structure, and other elements. Like classes in object-oriented programming, blueprints build a specialization hierarchy. This concept enables polymorphic instantiation of services (specialized services can be used as substitute of their more generic ancestors) and is the key concept supporting unconstrained reconfiguration at run-time.

Fig.4 illustrates an example of blueprint inheritance and decomposition hierarchy for a specific service named "Obstacle mapping", which is decomposed into three constituent service blueprints, "Distance sensor", Obstacle detection", and "Occupancy map generation". While in this example "Distance sensor" and "Obstacle map generation" are always basic service blueprint, "Obstacle detection" may be either a basic service blueprint or a service composition blueprint.

SoR is implemented by means of two basic components: *System Discovery* and *Reconfiguration* as illustrated in Fig.5. The availability of service instances is supervised via the *System Discovery* component, which allows all the available

Design-time Specification of DMSC in Support of Run-Time Safety Assessment   57

services to register themselves as available services at run-time and manages related information such as the corresponding service blueprint. The *Reconfiguration* component uses this information to create new configurations by traversing top-down through the service blueprints and instantiating service compositions and invoking available instances of basic services.

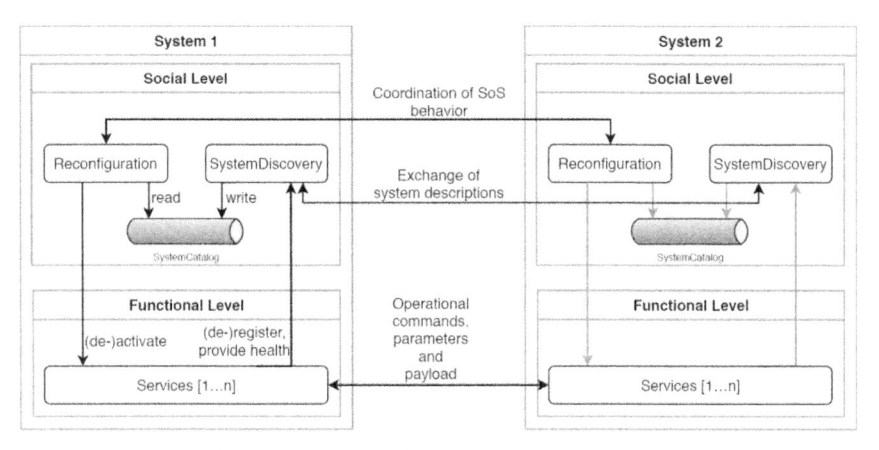

Fig. 5. The SoA-based concept of Service-oriented Reconfiguration (SoR)

In the basic SoR approach, this results in creating a set of possible configurations, from which the most suitable configuration is chosen based on evaluation of performance functions. For safety-related applications, one needs to ensure that only configurations are chosen for which their safety is assured. To achieve this, the DMSC concept proposed in this paper enables run-time creation and evaluation of SCs for each of the possible configurations.

3 Related Work

Several state-of-the-art approaches propose the automated development of SC based on the automated instantiation of argument patterns. An SC pattern specifies an abstract, reusable structure of a successful argumentation structure, containing placeholders for system-specific information, which can be filled in later, during pattern instantiation, i.e., during the usage of the pattern in the argument of a certain system (Kelly and McDermid 1998).

Denney and Pai provide formal semantics for creating SC patterns within GSN models, clarifying their restrictions and specifying a generic data model and pattern instantiation algorithm (Denney and Pai 2013). Following their work, they extended patterns with pattern metadata (Denney and Pai 2015), to capture the notion of tracing between pattern elements, e.g., informal claims. They further proposed formal foundations for composing different GSN arguments developed by patterns instantiation (Denney and Pai 2016). They defined arbitrary patterns composition, by taking the union of all links in the respective patterns, using shared identifiers as the points at which to join.

Finally, they implement their developed concepts in AdvoCATE (Denney and Pai 2018), a modelling tool which provides tool support for GSN-based SC pattern instantiation. However, the pattern instantiation is not a fully automated activity due to the usage of instantiated data table, which is manually specified by the safety engineer at design-time. Despite the provided formalization, they do not define algorithmic checks for the evaluation of composed patterns at run-time.

Šljivo et al. (Šljivo, et al. 2020) propose extended design pattern templates with contractual specifications, providing clear understanding of designed patterns compatibility with a given system environment, checking whether they fulfil the guaranteed safety claims. In particular, they define *PatternAssumptions*, representing conditions that shall be met for the correct usage of the design pattern, while the *PatternGuarantees* represent the conditions that the correct application of the pattern yields. The approach is implemented within the AMASS platform (De La Vara, et al. 2020), where contract-based reasoning via OCRA is enabled. Nevertheless, their approach has not been exploited for the dynamic reconstruction of SC.

Vierhauser et al. (Vierhauser, et al. 2021) introduce a mechanism for unmanned aerial vehicles (UAV) based on composable Safety Assurance Case (SAC). Their system assurance case is composed of: 1) an Infrastructure Safety Assurance Case (ISAC), which argues about the satisfaction of infrastructure-level safety goals, and 2) Pluggable Safety Assurance Cases (pSACs), which specify the safe operation of individual systems within a Complex open and adaptive system. They extend GSN with interlock points to dynamically plug at run-time sub-trees of pSAC to ISAC's interlock points. By combining their approach to monitoring methods at run-time, they check the validity of the entire SC. Their method similarly supports the idea of module assembly in SC for complex open and adaptive systems considering operational data. Nonetheless, they do not address the re-evaluation of composed modules patterns dynamically at run-time.

As explained earlier in this paper, DSC have been proposed to support the re-evaluation of the validity of design-time safety assumptions at run-time (Denney, Pai and Habli 2015), (Asaadi, et al. 2020). To this end, Denney and Pai propose that the SC is machine-comprehensible and hence, formalized.

Calinescu et al. introduced ENTRUST (Calinescu, et al. 2018), an end-to-end methodology for the engineering of trustworthy self-adaptive software systems, also implementing the DSC approach. They propose the development of a system safety assurance using SC patterns, which are partially instantiated with placeholders for the assurance evidence that cannot be obtained until the uncertainties associated with the system are resolved at run-time. One proposed pattern argues about the satisfaction of a set of safety requirements by the current system configuration. Given a system reconfiguration, ENTRUST proposes the re-verification of all safety requirements and, based on the obtained verification evidence, the re-instantiation of the respective SC pattern. Still, the re-verification of all requirements is time-consuming and not in line with the need for the rapid assurance of the new configuration required by open adaptive complex systems.

Cheng et al. (Cheng, et al. 2020) introduced the AC-ROS approach, which enables Robot Operating System (ROS) based platforms to conform to GSN models at run-time, thereby assuring that the system continues to satisfy its safety requirements after system reconfiguration. Despite this approach provid first steps towards DSC, its scope is limited to ROS-based systems.

Although state-of-the-art approaches enable initial support for dynamic SC management, they do not provide specific guidelines to develop SC elements such as modules and patterns to enable their composition in line with structural changes of open complex and adaptive systems.

4 DMSC Specifications

In this section, we propose an automated approach for constructing the SC of open complex and adaptive systems in a modular manner, reflecting the current system configuration.

4.1 General concepts

First, to allow the automated construction of SC, there is a need for a formal specification of SC, following a certain structure, with certain semantics. On the one hand, GSN is one of the most frequently used notations for structuring SCs. On the other hand, Yan et al. reviewed the current Model Based Engineering (MBE) techniques for generating SCs and they suggested that the SACM metamodel can support automatic SC generation, which they claim reduces the workload and the potential for errors, and supports SC evolution along with the system development (Wei, et al. 2019) , (Yan, Foster and Habli 2021). Consequently, for the specification of DMSC, we use the GSN/SACM metamodel, presented in Section 2.1.

The DMSC approach differentiates between the SC construction at design-time and at run-time. It proposes module-based assembly of SC, in correspondence to the open complex and adaptive systems reconfiguration. In other words, for each new configuration, the system SC is re-constructed based on automated composition of SC modules.

To facilitate this, the composition of SC modules must mirror the architecture of the open complex adaptive system, i.e., the composition of the network of services. For each service and each safety property that the service needs to satisfy, one module is constructed. Goal decomposition within such a SC module corresponds to the service composition specified by the blueprint implemented by the respective service. Such a module has a direct trace link to the related blueprint.

We categorize SC modules based on the service blueprint it addresses as follows:

- **Composed module**: specifies the safety argument corresponding to a service composition. The argumentation within such module can be further developed in other basic and/or combined modules, arguing about the composed services.
- **Basic module**: specifies the safety argument associated to a basic service.

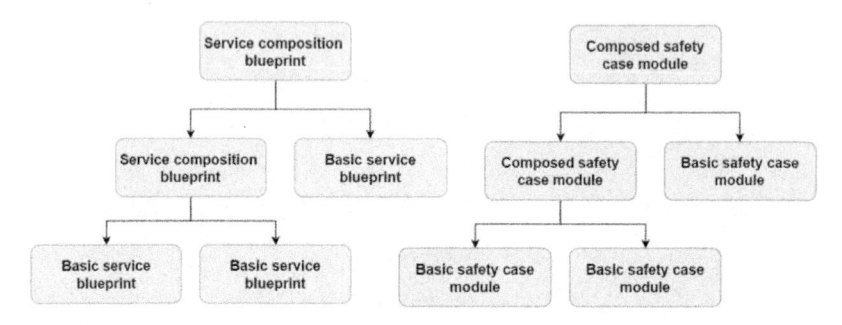

Fig. 6. Mapping between safety case and service blueprint architecture

Fig. 6 describes the mapping between SoR and SC architecture and Fig. 7 illustrates an overview of the proposed approach for DMSC development.

In the next subsections, we describe how to develop the SC of a SoR-able system, by differentiating the steps that need to be taken at design-time and the one to be taken at run-time.

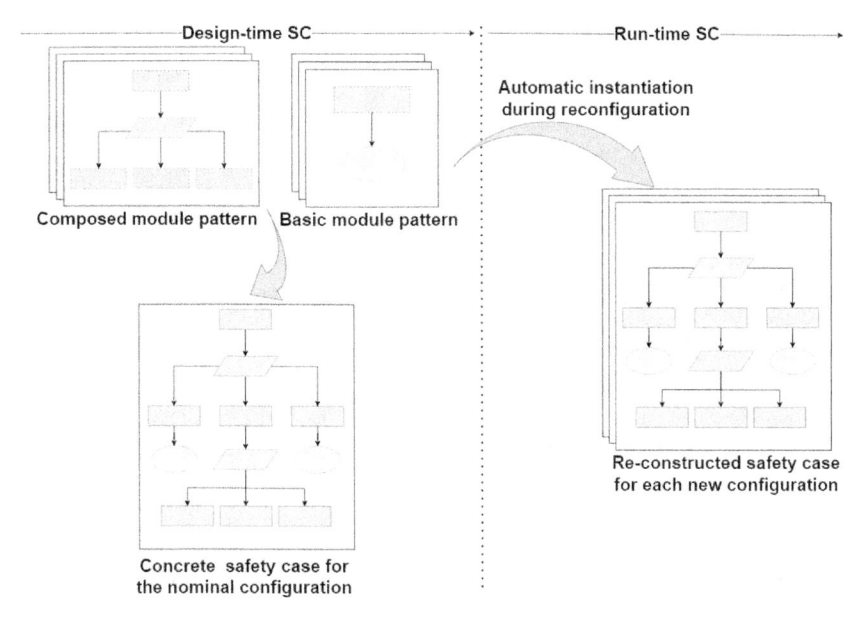

Fig. 7. General overview of the DMSC approach application at design-time and run-time

4.2 Design-time SC specifications

In order to support the generation of SC modules, we use SC patterns. Each pattern and the modules instantiating that pattern provide a trace link to a service blueprint, and, respectively, to the available services. Such a mapping enables automatic module composition respective to the service composition within each configuration. Consequently, a concrete SC can be re-constructed and instantiated at run-time for each possible system configuration.

First, to ease the specification of SC modules arguing about the assurance of basic services, for each property type, we propose a SC pattern, which can be instantiated for each available basic service. Such a module entails a top-level *Goal* about the satisfaction of one safety property (e.g., the failure rate of the addressed service). This module provides a fully developed argumentation, namely they reference the evidence on which the argumentation is based. We assume that the argumentation about safety properties of basic services is not subject to change during reconfiguration (since the reconfiguration, as realized in SoR, affects structure only), so that the design-time argumentation related to basic services remains untouched and valid during run-time reconfiguration.

Similarly, we propose a SC pattern for each composed module, which addresses the service composition blueprint and each safety property that needs to be demonstrated. In comparison to the basic modules, the argumentation within composed modules is not completely developed, but it is based on argumentation about the safety assurance of the composed services. To achieve this, these patterns entail *Away Goals*, each pointing at instantiation to a SC module arguing about the safety assurance of a constituent service of the service composition. Further, these patterns entail a strategy explicitly indicating how different *Away Goals* support the top-level *Goal*. The top-level *Goal* guarantees the satisfaction of a certain safety property only if the assumptions correspondent to these properties, which are supported by *Away Goals*, are valid. Such a strategy may specify a safety property as a function that takes the qualitative or quantitative guarantees of the safety properties corresponding to the constituent services as inputs and combines them to compute the valid result for the service composition blueprint. For instance, for the failure rate as a safety property, assuming the service composition blueprints are decomposed to a series of basic service blueprints, the top-level *Goal* guarantees will be computed as the sum of all failure rates provided by the pointed *Away Goals* (see Fig.8).

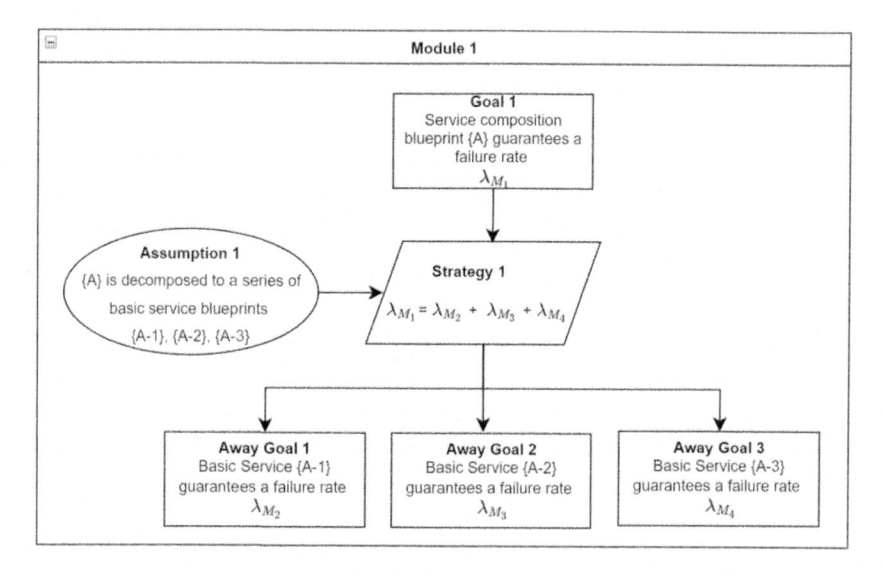

Fig. 8. Safety property function embedded into strategy to calculate the provided guarantees from basic service blueprints

For open complex adaptive systems, constructing SC modules following the previous specification leads to creation of the SC structure. The key information to determine which basic and/or composed modules can be assembled later at runtime are the instantiation mapping data linking the SC to the service blueprint architecture.

Design-time Specification of DMSC in Support of Run-Time Safety Assessment   63

At design-time, the SC of the initial nominal configuration is developed by connecting the SC module scoping a nominal service composition with basic and/or composed SC modules via *Away Goals*.

We extend the GSN/SACM metamodel having a claim associated to a certain type of service composition blueprint. This Away *Goal* is supposed to point to a *Goal,* which supports a claim associated to a certain type of basic and/or service composition blueprint. This leads to automatic instantiation of different *Away Goal*s along with service composition blueprint instantiation. Fig. 9 presents an overview of our proposed extension for GSN/SACM metamodel.

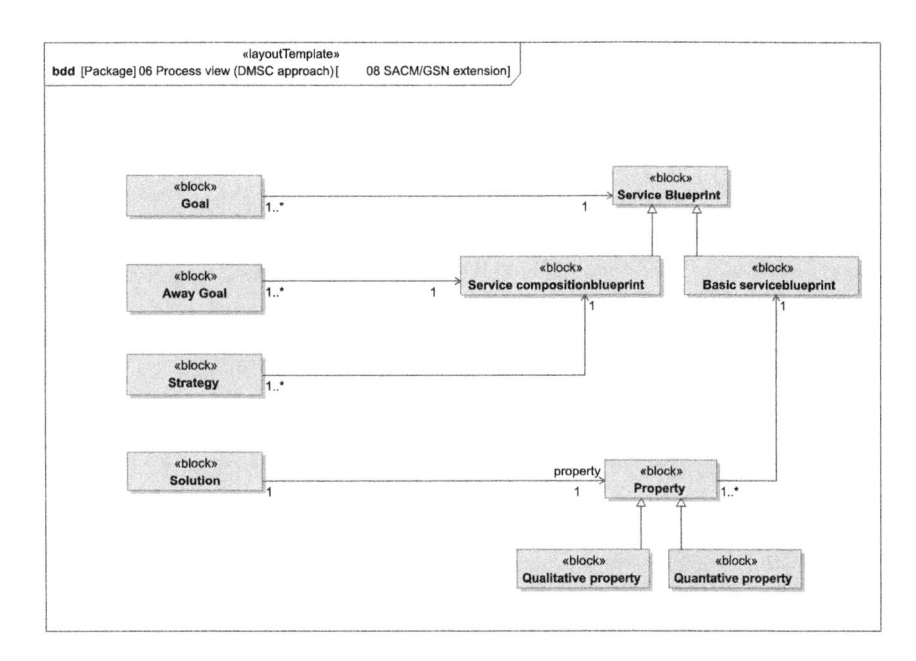

Fig. 9. The extended GSN/SACM metamodel supporting DMSC

To this end, we need to be able to specify a direct trace link between a *Goal* and one of the available service blueprints. To enable the specification of such direct trace links, we make use of the modelling capabilities offered by SACM (Selviandro, Hawkins and Habli 2020). In the following, we explain how SACM supports the specification of direct trace links, leaving out any other modelling concepts that are not relevant for this specification.

According to the GSN/SACM, all GSN constructs extend the `ArtifactElement` SACM class, which extends the `ModelElement` SACM class. Consequently, inheriting from the `ModelElement` SACM class, any GSN construct has a `description`. The GSN/SACM relationship is illustrated in (ACWG-GSN-MM 2021).

The `description` specifies the claim of a GSN construct in `MultiLang-String`, i.e., different languages (e.g., various natural languages such as English or German, or more formal languages such as Linear Temporal Logic). Further, a `description` may entail one or more `Terms`. A `Term` actually specifies a direct trace link or a placeholder for a direct trace link to a certain type of artefact. Each `Term` has an `externalReference` to a referenced artefact (i.e., models and model elements) of the type specified by the type attribute. Terms with empty `externalReference` can be used as to-be-instantiated parameters of abstract claims in parametrized safety case patterns (Matsuno and Taguchi 2011) i.e., placeholders for concrete trace links. Consequently, in our approach we use `Terms` in the claims of SC elements to establish mappings between the SC model and the service blueprint models.

4.1 Run-time SC construction and evaluation

For the purpose of run-time re-evaluation of each composed SC module, we formalise the design-time SC to be machine readable using the `SysML v2.0` textual notation[2], with some minor extensions. Once the reconfiguration is triggered, the SC modules will be re-assembled following the provided mapping between SC and blueprint architecture at design-time, which facilitates the instantiation of blueprints alongside with their SC modules per each new created configuration.

Further, for verifying the validity of new constructed SC, we check assume/guarantee relations, where the assumptions specify safety-related properties assumed in the current module that are expected to be demonstrated as valid by the modules pointed to by the *Away Goals*, and the guarantees specify safety-related properties that are demonstrated by the current module, given the satisfaction of its assumptions. The SC evaluation algorithm re-assesses each possible new configuration by verifying the assume/guarantee relation through traversing in the new composed SC modules. Eventually, if no violation is identified in the guarantees, the configuration is assessed as valid. Consequently, the configuration becomes part of the set of valid configurations, from which the SoR *Reconfiguration* component selects the most suitable configuration as the target configuration to be implemented.

---

[2] https://github.com/Systems-Modeling/SysML-v2-Release

5 Tool Support and Example

In our proposed approach, part of the development and assurance artefacts is done at design-time, such as the specification of service blueprints, safety case patterns, and SC modules. Hence, tool support for the specification of these artefacts would be beneficial.

5.1 Tool implementation

In this section, we discuss how we extended the FormAl SpecificaTion Environment (FASTEN)[3] in order to offer the needed tool support. FASTEN is an open-source environment for the specification, verification and assurance of safety critical systems. One characteristic of FASTEN is that it allows the deep integration between models of different aspects of the system (Ratiu, et al. 2021), e.g., a safety case model and the system architecture. FASTEN is built on JetBrains Meta Programming System (MPS), which is an open-source language workbench that targets Domain-specific Languages (DSLs). In contrast to general-purpose languages (GPLs), DSLs support the specification of systems in languages that directly use the concepts and logic from a specific application domain. Basically, FASTEN is a stack of DSLs, easily extensible via the specification of new DSLs. Among others, FASTEN has DSLs for the specification of system architecture, GSN-based safety cases, and of GSN-based SC patterns.

To enable the modelling of service blueprints presented in Subsection 2.2, and the mapping between SC and service blueprint models, we extend the FASTEN platform with a new stack of DSLs – FASTEN.DMSC. To enable the specification of direct trace links from safety case elements to the specified service blueprints, we extend the `IWord` interface from the FASTEN platform, which enables the specification of direct trace links from one model element to another. Examples of such trace links can be seen in Fig.6.

5.2 Example

We next illustrate how we apply the DMSC development guidelines, which are proposed in Section 4, to a simple, but clear example, while using FASTEN for the modelling activities. The example considers a scenario in a factory layout, where a group of two transport robots, robot R1 and robot R2, collaborate with each other.

---

[3] https://sites.google.com/site/fastenroot/home

We focus on the service composition blueprint of obstacle mapping – a service, which is provided by both robots from our example. In Fig.10 we show a screenshot from FASTEN tool, with the editor where we modelled the obstacle mapping service blueprint. According to the blueprint, an obstacle mapping service is composed of three other services, namely the distance sensor service, the obstacle detection service, and the occupancy map generator service.

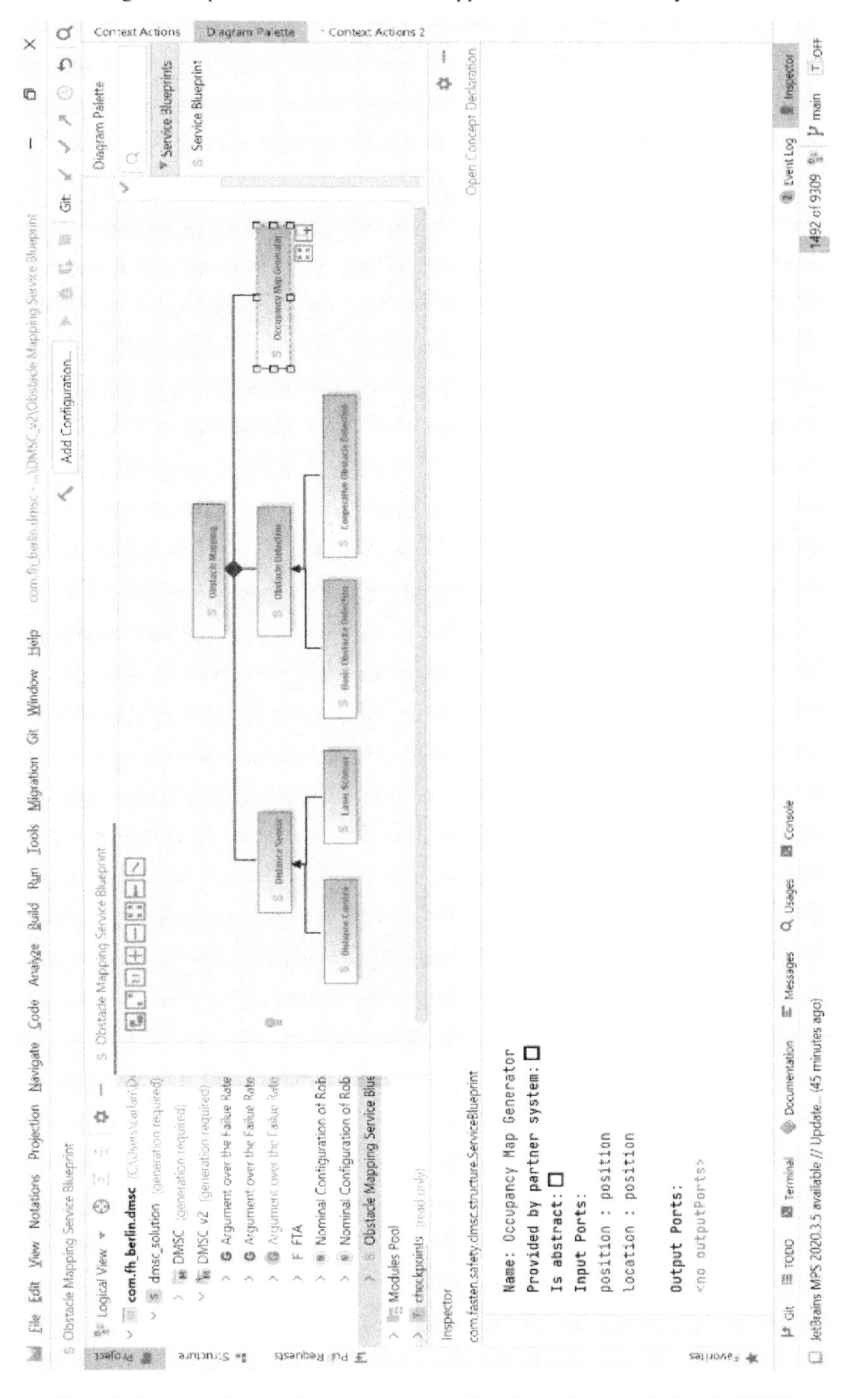

Design-time Specification of DMSC in Support of Run-Time Safety Assessment 67

Fig. 10. The obstacle mapping service composition blueprint modelled in FASTEN

Next, we present the steps taken for the modelling of a modular SC arguing about the safety assurance of the obstacle mapping service provided by robot R1. The SC justifies the safety of associated failure rate to this service being sufficiently low, where the failure rate of the composed service is computed from the failure rates of the composing services. For simplification purposes, we assume that the failure rates of each service are independent.

As a first step, we model in FASTEN an SC pattern arguing about the failure rate met by a given basic service (see Fig.11.a) The top-level goal (G1) of this pattern has a placeholder for a direct trace link to the addressed blueprint, using an extension of the `IWord` interface. The argument is supported by the results of a Fault Tree Analysis (FTA). Based on this pattern, at design-time, we create a set of SC basic modules, each corresponding to an available basic service. For our example, we create the SC modules corresponding to the available basic services of both robot R1 and robot R2: the basic distance sensor, the basic obstacle detection, and the basic occupancy map generation. Further, we also create a pattern arguing about the fact that the failure rate met by a service composition is sufficiently low (see Fig.11.b).

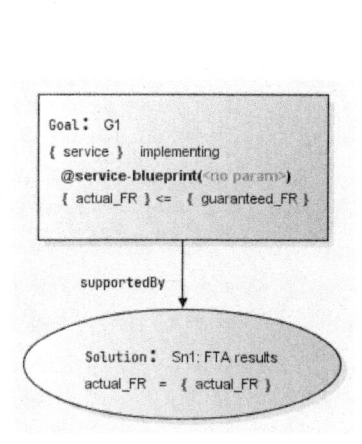

Fig. 11a Safety case pattern for arguing about the sufficiency of the failure rate of a basic service

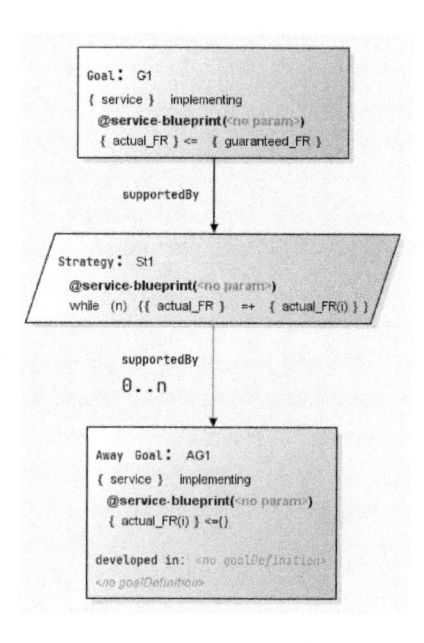

Fig. 11b Safety case pattern for arguing about the sufficiency of the failure rate of a service composition

Next, we model a SC module developed from a partially instantiated pattern by arguing that a system implementing the composed obstacle mapping service has

Design-time Specification of DMSC in Support of Run-Time Safety Assessment 69

a failure rate that is sufficiently low. The structure of this SC module mirrors the composition of services presented by the service blueprint model (see Fig.12).

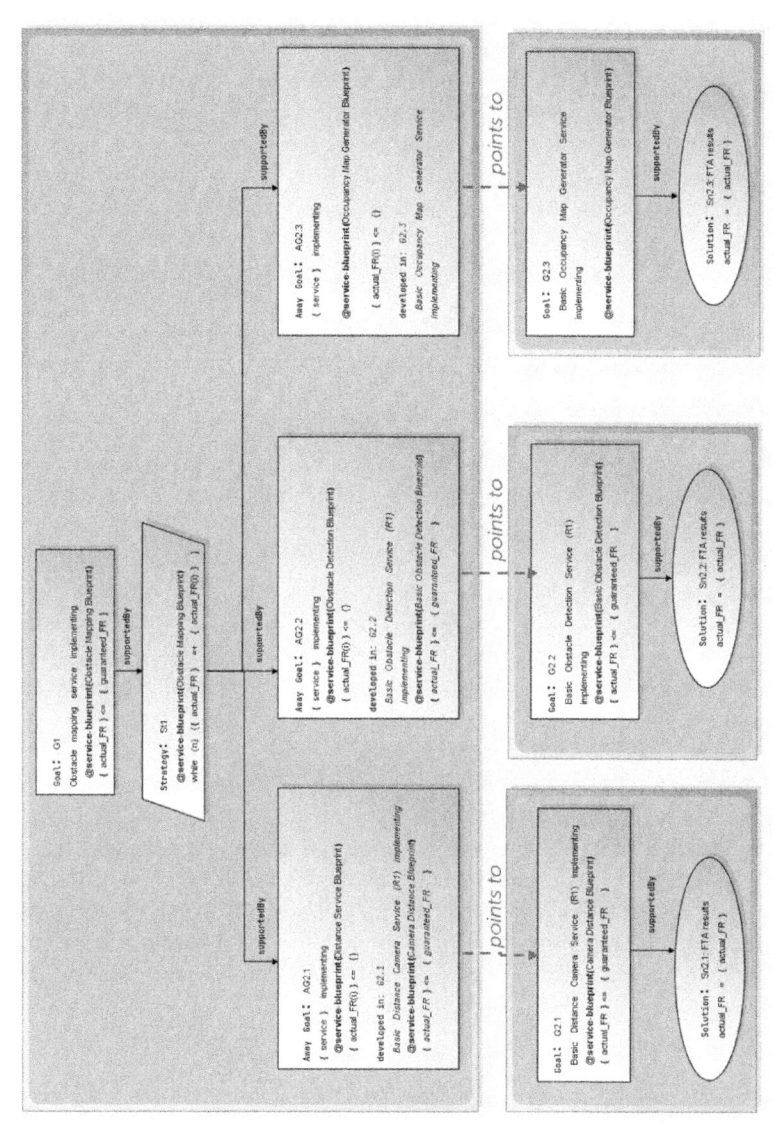

Fig. 12. The concrete safety case for the nominal configuration instantiated from the obstacle mapping service composition blueprint at design-time

As discussed in Section 4, considering the nominal configuration, the top-level safety *Goal* is instantiated at design-time with a reference to the composed service blueprint. Since the obstacle mapping service is composed by basic and/or service compositions blueprints, the satisfaction of the top-level *Goal* is demonstrated by arguing that the failure rates of the components implementing the service blueprints are sufficiently low. Therefore, the argumentation within this SC pattern is supported by *Away Goals*, which, after instantiation, point to SC modules arguing about the failure rates met by the services composing the obstacle mapping service. At design time, the *Away Goals* are instantiated considering the nominal configuration. At run-time, whenever a reconfiguration occurs, whereas the instantiated top-level *Goal* does not undergo any other changes, the *Away Goals* are to be re-instantiated, i.e., they will point to different SC modules, depending on the chosen composing services in the new configuration.

In Fig.12, we show how the SC for the nominal configuration is modelled in FASTEN, based on the instantiated patterns. The nominal configuration of robot R1 implements the obstacle mapping service by composing three basic services for distance sensor, occupancy map generator and obstacle detection service. Therefore, the *Away Goals AG2.1, AG2.2, AG2.3* in the safety case arguing about the failure rate of the obstacle mapping service point to *G2.1, G2.2* and *G2.3*, which argue about the fact that the failure rates associated to these basic services are sufficiently low.

Once we create the design-time patterns and modules, we formalized the modules using `SysML v2.0` textual notation for the SC instantiation and evaluation at run-time. As a reconfiguration is triggered, a new SC fragment composing the modules scoping the composing services is created. For all the new possible configurations, a re-evaluation will be done verifying the assume/guarantee relations between the obstacle mapping module and the distance sensor basic and/or composed modules.

In our example, for the obstacle mapping modules pattern, we assume the option to instantiate either basic or cooperative services for the distance sensor, obstacle detection, and occupancy map generation. According to the combination formula (Cameron 1994), the three options out of six available services result in 20 possible configurations – which is a considerable number for a such simple example. Whilst concrete safety cases for these 20 configurations could be created and evaluated at design-time, this is impossible for any complex open adaptive system of meaningful size and complexity. Here, the number of possible configurations easily reaches thousands.

6 Conclusion and future work

In this paper, we continue our ongoing line of work on developing DMSC to support SoR within the context of complex open and adaptive systems. In particular, we describe how to develop a design-time SC in a structured manner to facilitate safety assessment during reconfiguration at run-time. To this end, we outline the guidelines for the design-time specification of DMSC, while also using SC patterns. Further, we enable the co-evolution between system and safety architectures by defining trace links between the service blueprint architecture in SoR and the SC hierarchy established using the DMSC method and patterns. This also facilitates more purposeful and systematic SC maintenance by restricting the propagation of change impact only to certain SC modules.

Together with the proposed design-time SC, we additionally propose the formalization of SC modules with the purpose of automated pattern instantiation and re-construction of a concrete SC at run-time for each new created configuration. The SC automatically created at run-time can be evaluated by validating the assume/guarantee relations between modules and assuring the module composition. Further, we provide tool-support for our proposed safety case development guidelines by extending FASTEN – a system and safety engineering platform with capabilities for modelling service blueprints, service-oriented architectures based on those blueprints and DMSCs. Finally, we applied the proposed DMSC specification guidelines to an example from the smart factory domain.

In this paper, we propose the development of a DMSC at run-time, via the composition of SC modules specified at design-time. As a next step, we will elaborate on how to define formalised assume/guarantee contracts for each SC module and we will propose an automatic analysis of the compatibility of these contracts.

Acknowledgments Barbara Gallina is partially supported by the by Sweden's Knowledge Foundation via the SACSys (Safe and Secure Adaptive Collaborative Systems) project. Elham Mirzaei and Carsten Thomas are supported by the German Ministry for Education and Research in frame of the ITEA3 research project CyberFactory#1[4] under funding ID 01IS18061D.

References

ACWG-GSN. 2021. "Goal Structuring Notation Community Standard." Vers. 3. Safety-Critical Systems Club (SCSC) Assurance Case Working Group (ACWG). https://scsc.uk/r141C:1?t=1.

---

[4] https://www.cyberfactory-1.org/home/

ACWG-GSN-EX. 2021. "GSN-SACM Argumentation Example." Vers. 2.1. Safety-Critical Systems Club (SCSC) Assurance Case Working Group (ACWG). https://scsc.uk/file/gc/GSN2SACM_examples-1084.pdf.

ACWG-GSN-MM. 2021. "GSN Metamodel Specification." Vers. 2.2. Safety-Critical Systems Club (SCSC) Assurance Case Working Group (ACWG). https://scsc.uk/file/gc/GSN_metamodelV2-2-1210.pdf.

Asaadi, Erfan, Ewen Denney, Jonathan Menzies, Ganesh J Pai, and Dimo Petroff. 2020. "Dynamic Assurance Cases: A Pathway to Trusted Autonomy." *Computer* 53 (12): 35 - 46.

Boardman, John, and Brian J Sauser. 2006. "The Meaning of System of Systems." Edited by IEEE. *IEEE/SMC International Conference on System of Systems Engineering.* Los Angeles, CA, USA: IEEE. doi:10.1109/SYSOSE.2006.1652284.

Bradbury, Jeremy S, James R Cordy, Juergen Dingel, and Michel Wermelinger. 2004. "A survey of self-management in dynamic software architecture specifications." *WOSS '04: Proceedings of the 1st ACM SIGSOFT workshop on Self-managed systems.* New York, NY, USA: Association for Computing Machinery. 28–33.

Calinescu, Radu, Danny Weyns, Simos Gerasimou, Muhammad Usman Iftikhar, Ibrahim Habli, and Tim Kelly. 2018. "Engineering Trustworthy Self-Adaptive Software with Dynamic Assurance Cases." *IEEE Transactions on Software Engineering* 44 (11): 1039 - 1069. doi:10.1109/TSE.2017.2738640.

Cameron, Peter J. 1994. *Combinatorics: Topics, Techniques, Algorithms.* Cambridge University Press.

CENELEC. 2007. *EN 50129: Railway applications - Communication, signalling and processing systems - Safety-related electronic systems for signalling.* Standard, International Electrotechnical Commission.

Cheng, Betty H C, Robert Jared Clark, Jonathon Emil Fleck, Michael Austin Langford, and Philip K McKinley. 2020. "AC-ROS: assurance case driven adaptation for the robot operating system." *MODELS '20: Proceedings of the 23rd ACM/IEEE International Conference on Model Driven Engineering Languages and Systems.* New York, NY, USA: Association for Computing Machinery. 102–113.

De La Vara, Jose Luis, Eugenio Parra, Alejandra Ruiz, and Barbara Gallina. 2020. "The AMASS Tool Platform: An innovative solution for assurance and certification of cyber-physical systems." *Joint Proceedings of REFSQ-2020 Workshops, Doctoral Symposium, Live Studies Track, and Poster Track co-located with the 26th International Conference on Requirements Engineering: Foundation for Software Quality (REFSQ 2020),.* Pisa, Italy: CEUR Workshop Proceedings, CEUR-WS.

Denney, Ewen, and Ganesh Pai. 2013. "A Formal Basis for Safety Case Patterns." In *Computer Safety, Reliability, and Security*, by Guiochet J., Kaâniche M. Bitsch F., 21-32. Berlin & Heidelberg, Germany: Springer.

Denney, Ewen, and Ganesh Pai. 2016. "Composition of Safety Argument Patterns." In *Computer Safety, Reliability, and Security. SAFECOMP*, by Guiochet J., Bitsch F. Skavhaug A., 51-63. Springer.

Denney, Ewen, and Ganesh Pai. 2015. *Safety Case Patterns: Theory and Applications.* Technical, NASA.

Denney, Ewen, and Ganesh Pai. 2018. "Tool support for assurance case development." *Automated Software Engineering* (Springer) 25 (3): 435-499.

Denney, Ewen, Ganesh Pai, and Ibrahim Habli. 2015. "Dynamic Safety Cases for Through-Life Safety Assurance." *IEEE/ACM 37th IEEE International Conference on Software Engineering.* Florence, Italy: IEEE. 587-590.

Eurocontrol. 2006. "Safety Case Development Manual, ed. 2.2." Eurocontrol (European Organisation for the Safety of Air Navigation).

Industrial Avionics Working Group. 2012. "Modular Software Safety Case Process Description." Accessed November 2021. https://www.amsderisc.com/wp-content/uploads/2013/01/MSSC_201_Issue_01_PD_2012_11_17.pdf.

ISO/IEC/IEEE. 2019. "ISO/IEC/IEEE 15026-1: Systems and software engineering - Systems and software assurance - Part 1: Concepts and vocabulary." Standard.

Kelly, Tim, and J McDermid. 1998. "Safety case patterns-reusing successful arguments." *IEE Colloquium on Understanding Patterns and Their Application to Systems Engineering (Digest No. 1998/308).* London, UK: IET. 3/1-3/9.

Kelly, Tim. 2003. "Managing Complex Safety Cases." In *Current Issues in Safety-Critical Systems*, by Anderson T. Redmill F., 99-115. London: Springer. doi:10.1007/978-1-4471-0653-1_6.

Kelly, Tim, and Simon Bates. 2005. "The Costs, Benefits, and Risks Associated With Pattern-Based and Modular Safety Case Development." *UK MoD Equipment Safety Assurance Symposium.*

Leveson, Nancy. 2020. "White Paper on Limitations of Safety Assurance and Goal Structuring Notation (GSN)." http://sunnyday.mit.edu/safety-assurance.pdf.

Matsuno, Yutaka, and Kenji Taguchi. 2011. "Parameterised Argument Structure for GSN Patterns." *11th International Conference on Quality Software.* Madrid, Spain: IEEE. 96-101. doi:10.1109/QSIC.2011.35.

Ministry of Defence. 2007. *Defence Standard 00-56: Safety Management Requirements for Defence Systems.* Standard, UK: Ministry of Defence.

Mirzaei, Elham, Carsten Thomas, and Mirko Conrad. 2020. "Safety Cases for Adaptive Systems of Systems: State of the Art and Current Challenges." *Workshops. EDCC 2020. Communications in Computer and Information Science.* Munich, Germany: Springer, Cham. 127-138. doi:10.1007/978-3-030-58462-7_11.

Object Management Group. 2020. "Structured Assurance Case Metamodel (SACM), Version 2.1." https://www.omg.org/spec/SACM/2.1/PDF.

Ratiu, Daniel, Arne Nordmann, Peter Munk, Carmen Carlan, and Markus Voelter. 2021. "FASTEN: An Extensible Platform to Experiment with Rigorous Modeling of Safety-Critical Systems." In *Domain-Specific Languages in Practice*, by Cicchetti A., Ciccozzi F., Pierantonio A. Bucchiarone A., 131-164. Springer, Cham. doi:doi.org/10.1007/978-3-030-73758-0_5.

Richardson, Chris. 2018. *Microservices Patterns: With Examples in Java.* New York: Manning Publications.

SAWG, SCSC. 2020. "SCSC Publications." Vers. V1.0. *SCSC.* Edited by Mike Parsons. Service Assurance Working Group (SAWG). February. Accessed November 2021. https://scsc.uk/scsc-156.

Selviandro, Nungki, Richard Hawkins, and Ibrahim Habli. 2020. "A Visual Notation for the Representation of Assurance Cases Using SACM." In *IMBSA 2020: Model-Based Safety and Assessment*, by Marc Zeller and Kai Höfig, 3-18. Springer. doi:10.1007/978-3-030-58920-2_1.

Siefke, Lennart, Volker Sommer, Björn Wudka, and Carsten Thomas. 2020. "Robotic Systems of Systems Based on a Decentralized Service-Oriented Architecture." *Robotics* 9 (4): 78. doi:10.3390/robotics9040078.

Šljivo, Irfan, Garazi Juez Uriagereka, Stefano Puri, and Barbara Gallina. 2020. "Guiding Assurance of Architectural Design Patterns for Critical Applications." *Journal of Systems Architecture* 110: 101765. doi:10.1016/j.sysarc.2020.101765.

Thomas, Carsten, Elham Mirzaei, Björn Wudka, Lennart Siefke, and Volker Sommer. 2021. *Service-Oriented Reconfiguration in Systems of Systems Assured by Dynamic Modular Safety Cases.* Vol. 1462, in *Dependable Computing - EDCC 2021 Workshops. EDCC 2021. Communications in Computer and Information Science,* by Rasmus Adler, Amel Bennaceur, Simon Burton, Amleto Di Salle, Nicola Nostro, Rasmus Løvenstein Olsen, Selma Saidi, Philipp Schleiss, Daniel Schneider and Hans-Peter Schwefel, 12-29. Munich, Germany: Springer. doi:10.1007/978-3-030-86507-8_2.

Vierhauser, Michael, Sean Bayley, Jane Wyngaard, Wandi Xiong, Jinghui Cheng, Joshua Huseman, Robyn Lutz, and Jane Cleland-Huang. 2021. "Interlocking Safety Cases for Unmanned Autonomous Systems in Shared Airspaces." Edited by IEEE. *IEEE Transactions on Software Engineering* 47 (5): 899-918. doi:10.1109/TSE.2019.2907595.

Wei, Ran, Tim P Kelly, Xiaotian Dai, Shuai Zhao, and Richard Hawkins. 2019. "Model based system assurance using the structured assurance case metamodel." *Journal of Systems and Software* 154: 211-233. doi:10.1016/j.jss.2019.05.013.

Wudka, Björn, Carsten Thomas, Lennart Siefke, and Volker Sommer. 2020. "A Reconfiguration Approach for Open Adaptive Systems-of-Systems." *2020 IEEE International Symposium on Software Reliability Engineering Workshops (ISSREW).* Coimbra, Portugal: IEEE. 219-222. doi:10.1109/ISSREW51248.2020.00076.

Yan, Fang, Simon Foster, and Ibrahim Habli. 2021. "Safety Case Generation by Model-based Engineering: State of the Art and a Proposal." *The Eleventh International Conference on Performance, Safety and Robustness in Complex Systems and Applications.* International Academy, Research, and Industry Association. https://eprints.whiterose.ac.uk/172352/.

Stories and narratives in safety engineering

Catherine Menon

University of Hertfordshire

Austen Rainer

Queens University Belfast

Abstract *The use of stories and narrative is widespread throughout safety engineering, from "war stories" to use cases. In this paper we consider the effectiveness of stories in modelling safety-critical systems and challenges. We present a discussion of how aspects of a story such as characterisation, narrative arc and setting can affect the extent to which it adequately illuminates a software engineering problem.*

1 Introduction

Storytelling is one of the oldest forms of human communication, from the narratives conveyed by primitive art and cave painting, to Greco-Roman myths, to the earliest known written story, the epic of Gilgamesh (Kovacs, 1989). Stories have been used to entertain, to moralise, to inform and – perhaps above all – to illuminate.

It is in this last capacity that we see the most significant use of stories in Science, Technology, Engineering and Mathematics (STEM) subjects, particularly engineering. Stories are used as abstractions of specific engineering problems: the Travelling Salesman (Robinson, 1949), the Dining Philosophers (Hoare, 1978), the Byzantine Generals (Lamport, 1982). In some cases the story becomes better known than its origin (Hoare's coinage of the phrase "dining philosophers", for example, is so strongly associated with the deadlock problem that Dijkstra's original "Dining Quintuple" (Dijkstra, 1987) is now largely forgotten – as are the previous usages of the term "dining philosophers" themselves (Acland, 1841, Martineau, 1838). Stories have a particular value in the pedagogy of computer science, as they contain elements of narrative and character which

© University of Hertfordshire 2022.
Published by the Safety-Critical Systems Club. All Rights Reserved.

students can recognise, and with which they identify. This initial sense of familiarity has been shown to decrease the perceived difficulty of subsequently understanding unfamiliar concepts such as algorithms, code and systems thinking (Parham-Mocello, Ernst and Erwig, 2019).

Although stories are an integral part of the way in which we teach, discuss and represent complex systems, historically there appears to have been very little thought given by engineers to the construction of *effective* stories. In this, of course, we refer only to fictional constructions and not the use of "war stories" or summaries drawn from accident reports. Fictional stories are unique in that their every property – characters, settings, plots – can be tailored to an effective representation of an abstract problem, and yet in many cases we still find that the properties of a constructed story are in conflict with the engineering problem which it seeks to illuminate.

In this paper we consider the effectiveness of stories in reasoning about and modelling safety-critical systems and situations. We present a discussion of how aspects of a story such as narrative, characters and plot can affect the extent to which it adequately illuminates a software engineering problem. We illustrate this with examples drawn from two specific stories representing underlying engineering problems: the Byzantine Generals Problem (Lamport, 1983) and the Dining Philosophers Problem (Hoare, 1978).

2 Stories and story characteristics

Much work already exists on the properties of stories (Yorke, 2014), the process of writing (King, 2000), (Lamott, 1994) and the philosophy behind storytelling as a concept (Bradbury, 1992). From these, we can extract some fundamental characteristics of a successful story.

In this paper we will use the term *primary characteristics* to refer to the essential properties which make a narrative a story (rather than an unrelated series of sentences): the presence of a protagonist, the presence of a desire or aim possessed by the protagonist, an obstacle (antagonist) leading to conflict, and a resolution in alignment with these properties. To be effective, a story must not only contain these elements, but they must be identifiable and understood by the reader, hence the question of the target readership becomes of paramount importance, and we return to this in Section 6. It may be stated that – outside of certain genres such as experimental fiction, which are unlikely to be relevant to engineering problems – possession of these primary characteristics is an essential requirement for all stories.

We will use the term *secondary characteristics,* by contrast, to refer to those aspects of a given specific story which correspond to the specific plot, setting and characters of a given story. For example, the secondary characteristics of the Byzantine Generals Problem (Section 3) include a city under siege, an unspecified number of generals, some traitors, etc. The secondary characteristics of the Dining Philosophers Problem (Section 4) include some stubborn and hungry philosophers, a single dining room and a limited supply of silverware; we suggest the conflict may safely be left as an exercise for the reader.

For a story to be effective, secondary characteristics must be consistent with each other and the assumed or explicit rules of the story's universe. That is, the characters should be of a kind which might credibly be found in that specific setting, and the plot should consist of events and choices which might credibly be made by these characters. Secondary characteristics are, in general, more varied than primary characteristics: we can find a story effective even where the characters act in surprising ways, or find themselves in an unusual setting – in fact, it is arguable from a position of literary criticism that a story without some element of surprise is the poorer for it. More pragmatically, in stories which represent an underlying engineering problem, an unusual protagonist (e.g. a general who dislikes killing), or an unexpected aspect of the setting (e.g. a world where safe passage is always given to messengers) can still be effective provided sufficient explanation and narrative effort is expended on ensuring the readers comprehend the "surprise" (Alwitt, 2002).

3 Byzantine Generals Problem

In this section we will consider a well-known story within the safety community: the Byzantine Generals Problem (BGP)[1]. This has the dubious virtue of representing a real flaw in some safety-critical systems: there has been at least one occurrence of a real-life Byzantine Generals Problem in the system failure of an Airbus A330 in 2020 (TTSB, 2021).

There are several presentations of the BGP, but in this paper we take what is perhaps the most commonly-cited within the safety engineering community: that described in (Lamport, 1982):

"We imagine that several divisions of the Byzantine army are camped outside an enemy city, each division commanded by its own general. The generals can communicate with one another only by messenger. After observing the enemy, they must decide upon a common plan of action. However, some of the generals may be traitors, trying to prevent the loyal generals from reaching agreement. The generals must have an algorithm to guarantee that the following two conditions are met:

---

[1] See also the paper by Dewi Daniels in this book, *"What do Byzantine Generals and Airbus Airliners Have in Common?"*

- BGP1: All loyal generals decide on the same plan of action [...]
- BGP2: A small number of traitors cannot cause the loyal generals to adopt a bad plan" (Lamport, 1982)

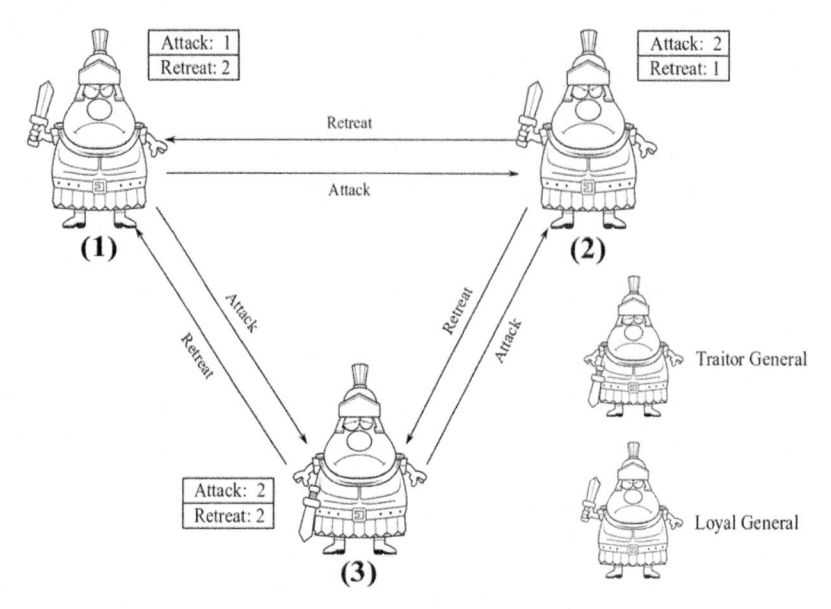

Fig. 1. Byzantine Generals Problem (Salimitari, 2020)

This is certainly a well-understood problem in safety engineering, and accurately represents the underlying challenge of communication between independent components of a system, some of which may not be reliable. Nevertheless, there are some amendments which could be made to the BGP as presented above to improve its accessibility as a story, and promote a deeper understanding of the problem outside of the safety engineering community.

We gave the BGP story to a group of professional writers, and after a single reading, a straw poll established that they presumed the following as part of the story:

Primary story characteristics:
- The generals want to agree on a method to defeat the city (e.g. by assault, sabotage, siege etc.)
- The traitors are trying to stop the generals defeating the city
- The generals will either defeat the city, be defeated in the attempt, or give up and (temporarily?) retreat

Secondary story characteristics:
- The generals are willing to incur some losses to defeat the city
- Messengers might be waylaid or put in danger by the traitors or the city

We emphasise that the writers were asked to react to this only as a story, rather than as a representation of an engineering problem. Moreover, it is also important to note that these readers do not have an engineering background or (as we established) prior knowledge of the BGP. They are therefore reading this story in the capacity of members of the lay public with particular experience in stories and their interpretation. This may go some way towards explaining some of the inconsistencies between the BGP as an engineering problem and the primary / secondary story characteristics deduced above.

Given these caveats, there are nevertheless some inconsistencies which make it clear that the BGP could be improved as a story, and hence as a means of communication about an engineering problem. In particular, none of the writers identified that that the generals' motivation is only to reach agreement, rather than to attack the city, and similarly they failed to identify that the traitors are trying to prevent agreement rather than defend the city. These misunderstandings of the story are discussed in more detail below.

3.1 BGP primary characteristics

A fundamental issue with the BGP – from a story rather than an engineering perspective – is the protagonists (the loyal generals) lack a desire consistent with what we know of their characters and the narrative. The Byzantine generals don't specifically desire to attack the city but instead merely want to reach agreement on what they should do next. While irreproachable as a representation of complex system communication, this choice is inconsistent with the characterisation of the generals (given the lack of any other information, readers assume generals are concerned primarily with conducting war and seeking victory rather than with anxiously ensuring that everybody agrees with everybody else).

Moreover, this desire is inconsistent with their previous assumed actions. The fact that the generals begin the story camping outside the city elicited from all readers the understanding that the generals desired to attack it. As a result, our straw poll readers all misunderstood the story in a way that would mischaracterise the engineering problem it represents: the BGP seeks to establish an algorithm for agreement amongst components, not an algorithm to obtain a particular specified decision outcome.

The second immediate problem with the BGP story is that the antagonists (traitor generals) also lack a desire consistent with their characterisation. They aren't seeking to save the city, but rather, desire only to prevent the loyal generals from agreeing. In this respect they are much more consistent with a "force of chaos" rather than a group of traitors (as an engineering-focused reader would expect, given the underlying engineering problem deals with system failure rather than malicious attack).

The third problem with the assumed primary characteristics of the BGP is the difference between the concept of a "plan of action" within the context of safety-critical system operation, and within the context of storytelling. In the safety-critical systems world, failure of the system components to come to an agreement represents a system failure with potentially catastrophic consequences. In the story world, if the generals fail to come to an agreement the reader's assumption is almost always that they will stay where they are. (They have not agreed to leave, attack, disband or take any other action, and within the story framework they must continue to exist somewhere: there is no "negative space" within the world of the story for them to vanish into).

Our straw poll of readers all interpreted "failure to come to an agreement" as a method of attack on the city. That is, the readers assumed that in the absence of an agreement the generals would stay where they were – and furthermore assumed that that constituted an attack by siege on a city that would eventually run out of resources. The readers therefore considered this a valid fallback option on the part of the generals, an implication which is missing within the safety-critical context. For a safety-critical system, of course, there is no concept of "still doing something" attached to a failure to decide.

We note that these problems stem from the story framing of the BGP, not the representation of the engineering problem. Specifically the setting of a city under siege and the generals as the protagonists lead readers to assumptions which align with stories about war, rather than stories about agreement, negotiation and diplomacy. We suggest that perhaps renaming this particular problem as the Byzantine Diplomats Problem might negate some of these misunderstandings!

3.2 BGP secondary characteristics

There are also some further concerns with the story presentation of the BGP in (Lamport, 1982). In particular, the story and engineering implementation are not separated: there is "story" information that can only be deduced by reading the engineering dissection of the problem. While this is understandable in the context of the original paper, it means that an inexperienced reader or member of the lay public has only an unclear – or worse, a misconceived – idea of the story, and hence the underlying engineering problem.

To take one example, the information given later in (Lamport, 1982) is that the generals must all receive the same information as each other, and moreover, that deciding on the same plan of action implies that the generals must have received the same information. However, the readers' assumptions in Section 3 about the secondary characteristics of the story are already in direct contradiction to this additional information. Specifically, all readers assumed that some generals might vote not to attack (either because it was personally detrimental for their troops, or because they had been given false information by a traitor general), but that the generals as a group could still agree on attacking even given disproportionate costs to a minority of their members. That is, as a group the generals are willing to incur some losses, and even those generals personally facing the losses would accept the plan.

However, if as given in (Lamport, 1982) there can be no such majority vote (i.e. the generals will only agree on a plan if they have been given the same information), then the reader's assumptions must be reassessed and corrected. That is, we are in a different story: the generals must all attack together if they are to have any hope of success, and any general who doesn't want to attack will refuse, thus ruining the plan. This is not the reader's stereotypical assumption of what coordinated army looks like and again we suggest an alternative: The Byzantine Loosely-Allied Tribes?

4 Dining Philosophers

We emphasise that the problems identified above with the BGP relate only to the story presentation, rather than the underlying engineering issue. To illustrate how an effective story can assist in our understanding of software, we next turn to another well-established story: the Dining Philosophers (DP):

> *"Five philosophers spend their lives thinking and eating. The philosophers share a common dining room where there is a circular table surrounded by five chairs, each belonging to one philosopher. In the centre of the table there is a large bowl of spaghetti, and the table is laid with five forks. On feeling hungry, a philosopher enters the dining room, sits in his own chair, and picks up the fork on the left of his place. Unfortunately, the spaghetti is so tangled that he needs to pick up and use the fork on his right as well. When he has finished, he puts down both forks and leaves the room"* (Hoare, 1978).

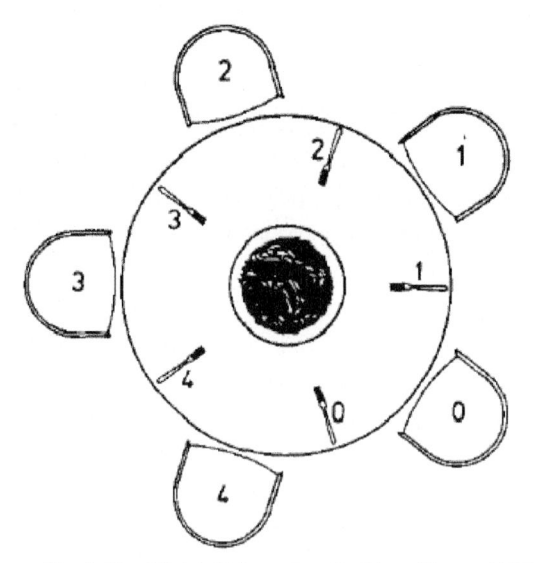

Fig. 2. The Dining Philosophers Problem (Hoare, 1978)

The DP as presented is an essentially absurdist situation: we are asked to accept that the philosophers will willingly starve should they not be able to obtain two forks with which to eat spaghetti. The inadequacy of their table manners notwithstanding, this scenario works as a story because of its internal consistency: a reader expects (stereotypical) philosophers to engage in nonsensical, overly-abstruse debate and to refuse all common-sense solutions. In other words the protagonists have a believable desire within the story world – to eat – believable characterisation, and the resultant conflict is in alignment with both.

It is also worth noting that the presentation of the DP in (Hoare, 1978) is a near-complete separation of implementation and "story". As such, it can be presented to non-technical readers in its entirety, without risking the contradictory assumptions which arise with the BGP.

We do, however, note one aspect of the BGP story presentation which equals – or betters – that of the DP: the story title. The original title of the Chinese Generals Problem (Lamport, 2021) is not only problematic from the perspective of racial prejudice, but it does not give readers any insight into why these generals are acting in a counter-intuitive and highly constrained manner. The translation to Byzantine generals, who might – in common with the rest of the Byzantine Empire – be expected to operate in a labyrinthine atmosphere of rules, constraints, plots and betrayal, accurately conveys the flavour of the story.

5 Discussion

As we emphasise in both Section 3 and Section 4, the critique of the stories presented here does not imply any criticism of the complexity, relevance or "worth" of the underlying engineering problems. Nevertheless, if stories are intended to convey an understanding of these problems – particularly to the lay public – it is worth examining why some are more effective than others.

Returning to the BGP of Section 3, we refer again to the observation in (Lamport, 2021) that the original problem related to Chinese generals. It is unclear from a modern position why these generals should be specifically Chinese – or, indeed, any other specific nationality, except Byzantine. Stories which appeal to stereotypes in this way often founder when such stereotypes become outdated or unknown. It is interesting that both the BGP in its current form and the DP (Section 4) hark back to relatively uncontroversial "ancient history" stereotypes of behaviour, rather than modern.

We suggest that it might also be instructive to consider whether the chosen story says something to readers about the importance of the problem. In general, the story chosen can encourage or discourage particular assumptions, or focus the reader's attention on a particular outcome. For example, an argument can be variously described as a war (with an implication of conflict, a winner and a loser) or as a dance (with an implication of decorum, collaboration and diplomacy). The construction and regulatory acceptance of a safety case argument is perhaps more akin to the latter in theory, and the former in practice!

In general, stories which postulate a potentially fatal outcome for one or more characters tend to receive more traction in safety discourse. The Trolley Problem (Foot, 1967) is an example: although this story largely does not capture the primary concerns of safety engineers, there are arguably few such engineers who have not referred to this during a discussion of autonomous vehicle safety. Similarly, the BGP is phrased as a story about "dramatic", high-worth events: war, invasion and treaties. We speculate that the story may have received less traction within the engineering community if the characters were of a very different kind (The Byzantine Schoolgirls Problem?).

6 Conclusions and further work

We acknowledge that seasoned engineers are unlikely to rely on the story of an engineering problem to provide them with their full understanding of it. However, safety-critical engineering is a discipline which relies on communication and public understanding of risk. Emerging technologies such as autonomous systems will require a greater degree of public acceptance, understanding and willing engagement (Information Commissioners Office and Turing Institute,

2020), and adequate communication – including in the form of accessible stories – is a necessary first step towards that.

From our preliminary steps in researching how the lay public interpret stories representing engineering problems, we have identified some axioms for constructing and communicating effective stories:

- The story must be an accurate model of the engineering problem. As with all models, omissions are inevitable, and are more tolerable than misrepresentations.
- The story and any engineering solution to the problem should not be mixed in the presentation
- Metaphors and assumptions within the story should be well-understood and, so far as possible, parallel the details of the engineering problem
- The setting and characterisation of the story should not rely on stereotypes which may be misunderstood, particularly where these are used to convey information about the underlying engineering problem
- The story must contain an element of drama, such as the potential for a fatal outcome for one or more characters

As future work, we propose to validate and extend these axioms, moving toward a comprehensive and engineering-focused theory of story construction. Specifically, we propose to examine how these axioms relate to two key safety outputs: safety case reports and accident investigation reports. These must both provide a compelling, credible story which is sufficient to convince the reader that the system is adequately safe in its proposed context of use (safety case report), or that the sequence of events leading to an accident has been comprehensively analysed (accident investigation report).

We also propose to explore how information can be transmitted in the opposite direction: that is, how a reader's reaction to an inconsistent narrative, characterisation or setting can be used to identify those aspects of the underlying engineering problem which might be inadequately specified. One specific approach to this would be to empirically investigate the transformation of the BGP story from the story through safety requirements, software design, code, and then execution and evaluation. As part of this process we will seek to investigate how variances in the detail of the story affect the transformations between lifecycle artefacts, and consequently affect our modelling of safety-critical systems. We propose to use this investigation to create a re-telling of the BGP, in a form which addresses some of the potential drawbacks of the current story.

References

Acland, A. (1841) A Letter to the Right Reverend Fathers In God. Ancient and Modern Ways of Charity, British Critic and Quarterly Review, Vol 29, J.G.F & J. Rivington.
Alwitt, L. (2002) Maintaining Attention to a Narrative Event, Advances in Psychology Research, vol. 18, pp. 99–114.
Bradbury, R. (1992), Zen in the Art of Writing, Bantam.

Dijkstra, E. EWD-1000 (1987) E.W. Dijkstra Archive, https://www.cs.utexas.edu/users/EWD/ewd10xx/EWD1000.PDF

Foot, P. (1967) The Problem of Abortion and the Doctrine of the Double Effect, Oxford Review, Vol 5.

Hoare, C.A.R. (1978) Communicating Sequential Processes, Communications of the ACM, Vol 21, Issue 8.

Information Commissioners Office, Turing Institute (2020) Explaining Decisions Made With AI. https://ico.org.uk/media/for-organisations/guide-to-data-protection/key-dp-themes/explaining-decisions-made-with-artificial-intelligence-1-0.pdf, accessed October 2021.

King, S. (2000), On Writing: A Memoir of the Craft, Scribner.

Kovacs, M. (1989) The Epic of Gilgamesh, Stanford University Press.

Lamott, A. (1994) Bird by Bird, Anchor.

Lamport, L., Shostak, R., Pease, M. (1982) The Byzantine Generals Problem. ACM Transactions on Programming Languages and Systems, Vol 4, 382-401.

Lamport, L. (2021) My Writings http://lamport.azurewebsites.net/pubs/pubs.html#trans, accessed October 2021.

Martineau, H. (1838) Retrospect of Western Travel, Saunders & Otley.

Parham-Mocello, J., Ernst, S., Erwig, M. (2019) Story Programming: Explaining Computer Science Before Coding, Proceedings of the ACM Special Interest Group on Computer Science Edcation Technical Symposium.

Robinson, J. (1949) On the Hamiltonian Game (A Traveling Salesman Problem), Rand Corporation RM-303, https://www.rand.org/pubs/research_memoranda/RM303.html, accessed October 2021.

Salimitari, M., Chatterjee, M., Fallah, Y. (2020). A Survey on Consensus Methods in Blockchain for Resource-constrained IoT Networks. Internet of Things, Vol 11.

Taiwan Transportation Safety Board (2021), Major Transportation Occurrence Final Report: Airbus A330, TTSB-AOR-21-09-21

Yorke, J. (2014), Into The Woods: How Stories Work and Why We Tell them, Penguin.

Development of Rechargeable Electrical Energy Storage Systems for Automotive and Aviation

Paul Malcolm Darnell[1], Pavan Venkatesh Kumar[2]

3S Knowledge Limited

Warwick, UK[3]

Abstract *The transition to electrified propulsion systems (EPS) within the automotive sector is maturing, many manufacturers have Battery Electric Vehicles (BEV) within their product offering. The hazards associated with EPS technology, which comprises of a Rechargeable Electrical Energy Storage System (REESS), charging systems, electric drive systems, thermal management systems, DC/DC converter, and supervisory control systems, are well understood and technical safety concepts are in series production. The aviation industry is now in an intense research and development phase for new energy propulsion systems, including electrification. This paper explores the potential to safely expedite the transition to 'electrified' aviation through reapplication of automotive EPS technology, or whether EPS technology may be viably developed for use in both automotive and aviation domains. A REESS case study is utilised, comparisons between technical and process orientated safety regulations and standards are made, and technical solutions are developed utilising an exemplar 'automotive-aviation' process model. The study shows that, while some unique new work products will be required, automotive technology may be utilised within aviation, subject to review and revision to ensure compliance with aviation regulations. The study also identifies opportunities to enhance the development process through sharing of best practice between the automotive and aviation domains.*

---

[1] Paul Darnell BSc(Hon) CEng MIET Hon DSc paul@3sk.co.uk

[2] Pavan Kumar MSc pavan@3sk.co.uk

[3] Innovation Centre, Warwick Technology Park, Gallows Hill, Warwick, UK, CV34 6UW 3SK.CO.UK

3S Knowledge Limited 2022.
Published by the Safety-Critical Systems Club. All Rights Reserved.

1 Introduction

The application of Rechargeable Electrical Energy Storage Systems (REESS) within the automotive domains is now maturing and the associated safety hazards are well understood. Proven safety concepts are in series production across many manufacturers of Battery Electric Vehicles (BEV). The REESS is a complex system consisting of lithium-ion cells, cell modules, Battery Management System (BMS), connectors, and housing as shown in figure 1. The BMS includes multiple sensors and controllers to manage the health and performance of the battery.

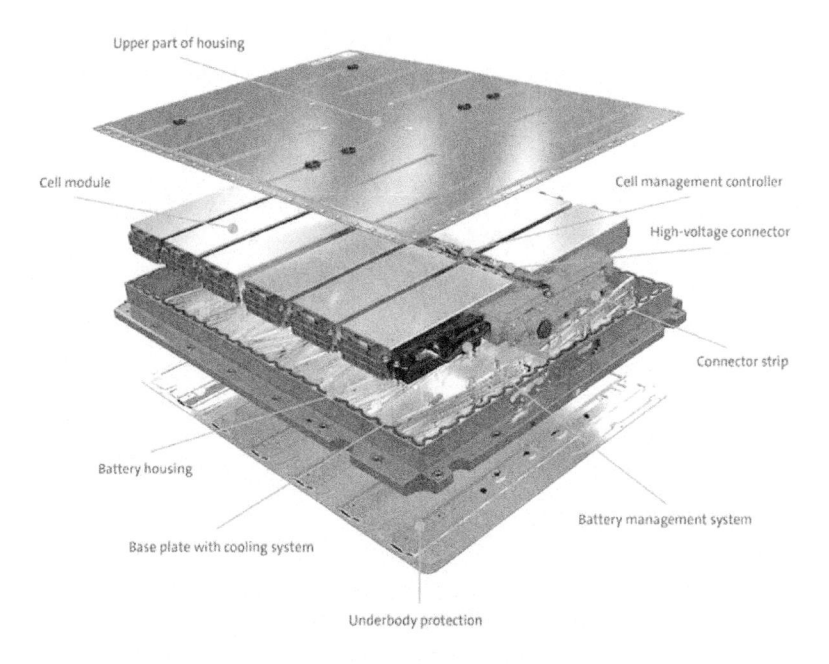

Fig. 1. Exploded view of REESS[1]

The aviation industry is now in an intense research and development phase of new energy propulsion systems, including electrification. This paper explores the potential to safely expedite this transition through reuse of REESS which were originally developed for the automotive domain for use in the aviation domain, or whether a REESS may be viably developed for use in both automotive and aviation domains. To enable robust conclusions to be established the following aspects are considered within this paper:

---

[1] https://www.volkswagen-newsroom.com/en/press-releases/in-brief-key-components-for-a-new-era-the-battery-system-5645

1. A review of domain specific regulations, standards, and guidelines.
2. Definition of a process model which is compliant with expectations of point 1 and supports the development for automotive and aviation technology.
3. Application of the process model for the development of a REESS system-level design (requirements and architecture) for both automotive and aviation domains.

The complete study of standards and regulations, definitions of development processes, and execution of these processes is a significant undertaking, and thus cannot be fully encompassed within the constraints of this paper. Therefore, we have strived to include the pertinent aspects of the study, and to provide useful insights and observations on whether a common reusable REESS for automotive and aviation is a viable proposition.

For this study, it is assumed that the REESS is the sole source of propulsion power for the vehicle or aircraft, hence the product is either a Battery Electric Vehicle or a Battery Electric Aircraft.

2 Safety regulations and standards

Numerous regulations, standards, and guidelines are applicable to the development of REESS. The scope of these documents related to both process and technical aspects, that is, they include details of the expected process activities, methods, work products, and the technical capabilities, constraints, test methods, and pass/fail criteria. The following sections review and compare these process and technical documents for both automotive and aviation domains.

2.1 Process documents

There are generic documents covering the development processes, methods, and work products expectations of automotive and aviation safety related systems, containing requirements, recommendations, and guidelines. These are not specific to the development of REESS but are applicable to REESS.

2.1.1 Automotive domain

ISO 26262 'Road vehicles - Functional safety' is the preeminent standard applied for the development of safety-critical systems within the automotive sector. Most organisations, whether Original Equipment Manufacturers (OEM) or Tier 'n' systems developers, reference and adhere to this standard across the globe.

ISO 26262 is one of many adaptations of IEC 61508 'Functional safety of electrical/electronic/programmable electronic safety-related systems' as illustrated in figure 2. The aviation standards precede the first release of IEC 61508 in 1998, and hence are not adaptations of IEC 61508.

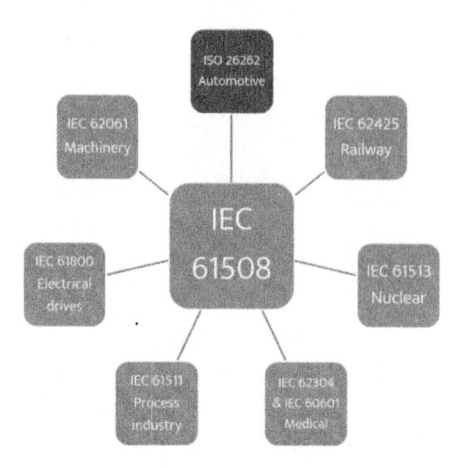

Fig. 2. Example of IEC 61508 derived standards

ISO 26262 encompasses the whole safety lifecycle of E/E systems, including their embedded software, spanning from concept phase, through system, hardware, and software development phases, followed by production, operation, servicing, and finally decommissioning of the product (figure 3).

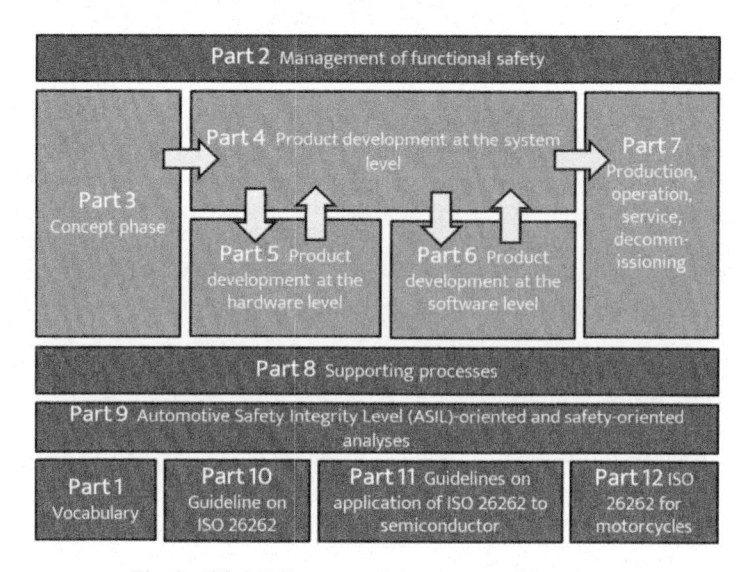

Fig. 3. ISO 26262 scope and development lifecycle

ISO 26262 contains both requirements and recommendations relating to the processes, methods, and work products. The rigour of these requirements corresponds to the Automotive Safety Integrity Level (ASIL) assigned to the elements of the system under development, and ranges from ASIL A (lower rigour) to ASIL D (higher rigour).

ASIL is determined from the severity, exposure, and controllability of each potential hazardous event which may occur in a range of operational scenarios, due to malfunctioning behaviour of the E/E system.

Severity: Estimate of the extent of harm to one or more individuals that can occur in a potentially hazardous situation

S0	S1	S2	S3
No injuries	Light and moderate injuries	Severe and life threatening injuries (survival probable)	Life-threatening injuries (survival uncertain), fatal injuries

Exposure: State of being in an operational situation that can be hazardous if coincident with the failure mode under analysis

E0	E1	E2	E3	E4
Incredible	Very low probability	Low probability	Medium probability	High probability

Controllability: Ability to avoid a specified harm or damage through the timely reactions of the persons involved

C0	C1	C2	C3
Controllable in general	Simply controllable	Normally controllable	Difficult to control or uncontrollable

Determination of ASIL

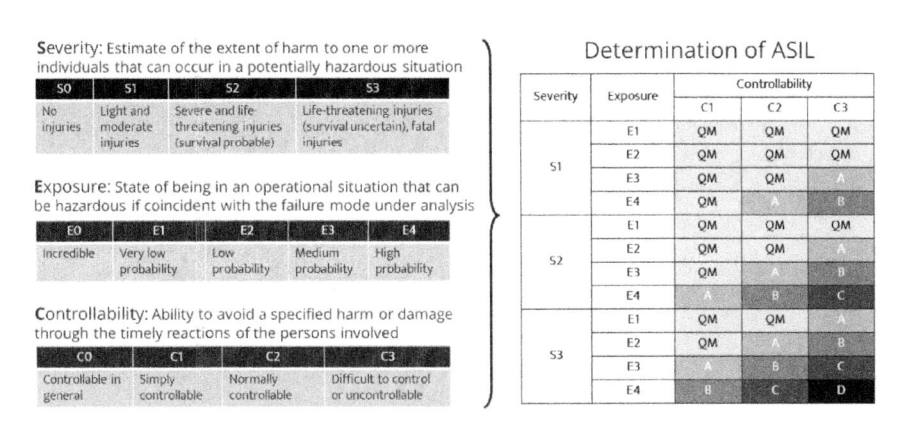

Severity	Exposure	Controllability		
		C1	C2	C3
S1	E1	QM	QM	QM
	E2	QM	QM	QM
	E3	QM	QM	A
	E4	QM	A	B
S2	E1	QM	QM	QM
	E2	QM	QM	A
	E3	QM	A	B
	E4	A	B	C
S3	E1	QM	QM	A
	E2	QM	A	B
	E3	A	B	C
	E4	B	C	D

Fig. 4. ASIL determination

As an example, if the severity of a hazard is classified as S3 (life threatening injuries), E4 (high probability of being in an operational scenario where these life-threatening injuries may occur), and C3 (difficult or not possible to control the product to avoid the injury), then ASIL D is determined[2].

If the risk of a hazard is evaluated as sufficiently low, then Quality Management (QM) is assigned. For QM, general high-quality processes are to be followed, such as those specified by ISO 9001 (Quality management systems) and Automotive SPICE[3].

Having a widely adopted standard, encompassing the whole safety lifecycle, enables a consistent and cohesive approach for the development of safety critical systems. It is worth noting that automotive certification authorities do not mandate compliance with ISO 26262, although ISO 26262 is generally recognised as state of art.

---

[2] SAE J9280 *Considerations for ISO 26262 ASIL Hazard Classification* provides additional information on risk assessment and ASIL determination

[3] http://www.automotivespice.com/ Automotive SPICE® is a registered trademark of the Verband der Automobilindustrie e.V. (VDA)

2.1.2 Aviation domain

There are multiple safety related documents within the aviation industry (figure 5) with two publishers, Radio Technical Committee for Aeronautics (RTCA) and SAE International (SAE).

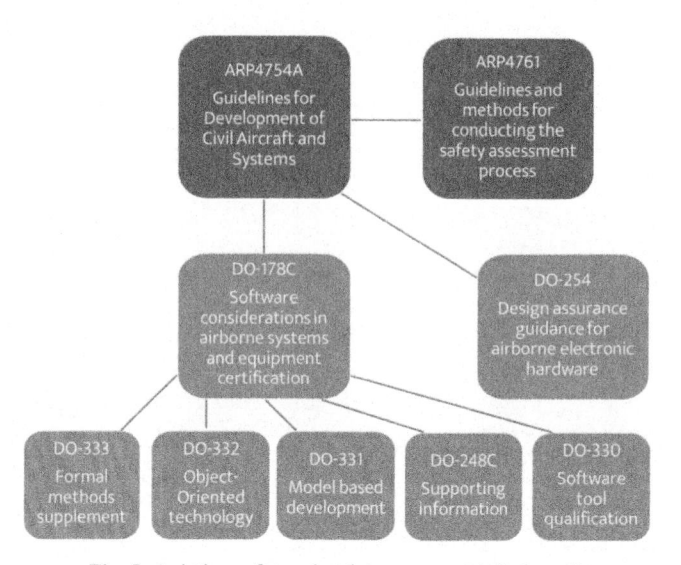

Fig. 5. Aviation safety related documents (civil aircraft)

RTCA first published documents associated with the software and hardware level development, DO-178 and DO-254 in 1992 and 2000 respectively, and SAE published the system level development documents ARP4754[4] and ARP4761[5] in 1996.

Since the focus of this paper is upon the system level development of the E/E elements within the REESS, the processes and methods of ARP4754A and ARP4761 will be considered in detail. Other documents will not be considered further herein. It must be stressed that to potentially reuse REESS from automotive to aviation, or develop REESS for both domains, adherence to all documents must be demonstrated.

The scope of ARP4754A covers the whole system development lifecycle (figure 6), and is not restricted to safety, albeit does include safety. ARP4754A includes 'integral' processes, which correspond to the 'supporting' processes of ISO 26262, although less extensive, and the ARP4754A planning (management) process is less prescriptive. Several practices within ISO 26262 could be beneficially applied to

---

[4] Guidelines for Development of Civil Aircraft and Systems, https://www.sae.org/standards/content/arp4754a/

[5] GUIDELINES AND METHODS FOR CONDUCTING THE SAFETY ASSESSMENT PROCESS ON CIVIL AIRBORNE SYSTEMS AND EQUIPMENT, https://www.sae.org/standards/content/arp4761/

aviation development, and these are explained throughout section 3 where each step in the development process is examined.

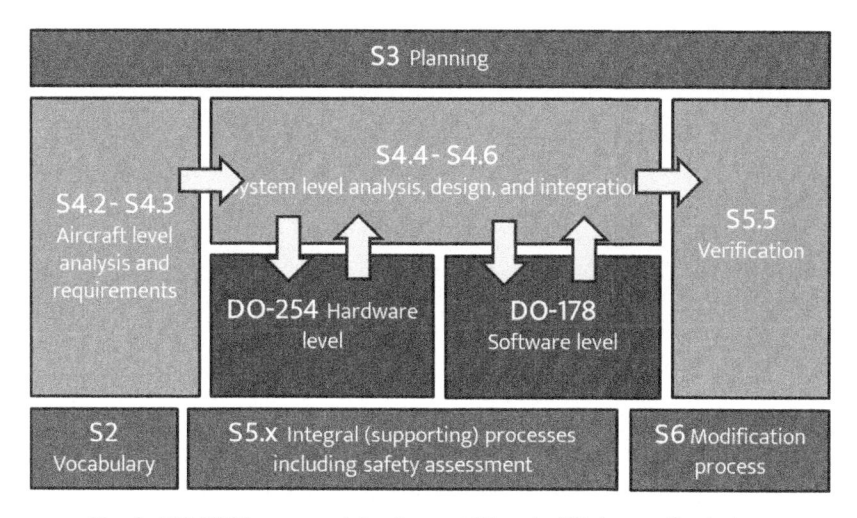

Fig. 6. ARP4754A scope and development lifecycle ('S' denotes Section)

ARP4761 provides substantial guidelines and methods for conducting the safety assessment processes identified within ARP4754A and summarised in figure 7. These guidelines are far more extensive than those of ISO 26262 Part 10 and may support those who are conducting safety analyses within the remit of ISO 26262.

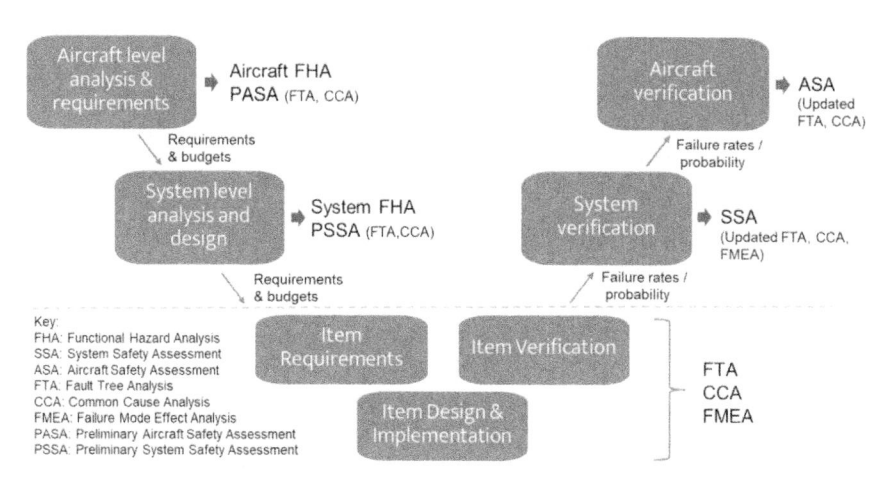

Fig. 7. ARP4754A Safety Assessment Process

The term 'item' is referenced throughout ISO 26262 and ARP4754A/ARP4761 and introduced in figure 7. It is worth noting a difference in the definition of the term 'item' between these documents:

- ISO 26262: *System(s)* that implement a function at the vehicle level
- ARP4754A: *Hardware* or *software element* with well-defined interfaces

As illustrated in figure 7, activities are conducted at aircraft level, system level, and item level. The activities may be iterated across different levels of system decomposition.

Fundamentally, the analysis methods are similar to those of ISO 26262, for example Fault Tree Analysis (FTA) and Common Cause Analysis (CCA), and despite the different product, the technical contents of the analyses for a REESS start to converge. Note that the necessary safety measures[6] may differ, and these are described in more detail in later sections.

During the Functional Hazard Analysis (FHA) the Development Assurance Level (DAL) (Table 1) is determined by analysing the severity of the failure effects for each function and considers the controllability of the failure by crew during each flight phase. Aviation flight phases are more constrained and defined compared to those of automotive (i.e. vehicle use cases and exposure).

Table 1. Design Assurance Level (DAL)

Classification	Description	DAL
Catastrophic	Failure may cause deaths, usually with loss of the airplane	A
Hazardous	Failure has a large negative impact on safety or performance, or reduces the ability of the crew to operate the aircraft due to physical distress or a higher workload, or causes serious or fatal injuries among the passengers	B
Major	Failure significantly reduces the safety margin or significantly increases crew workload. May result in passenger discomfort (or even minor injuries)	C
Minor	Failure slightly reduces the safety margin or slightly increases crew workload. Examples might include causing passenger inconvenience or a routine flight plan change	D
No Effect	Failure has no impact on safety, aircraft operation, or crew workload	E

In a similar manner as for ASIL, the assigned DAL correlates to the necessary level of process rigour applied during the development and the target failure budgets for random hardware failures. The target rates for both ASIL and DAL are shown in figure 8.

---

[6] Safety measures: Activities or technical solutions (safety mechanisms) to avoid or control systematic failures and to detect random hardware failures or control random hardware failures, or mitigate their effects (ISO 26262, 2018)

Target failure rates

	$<10^{-3}h^{-1}$	$<10^{-4}h^{-1}$	$<10^{-5}h^{-1}$	$<10^{-6}h^{-1}$	$<10^{-7}h^{-1}$	$<10^{-8}h^{-1}$	$<10^{-9}h^{-1}$
ASIL					B, C	D	
DAL	D		C		B		A

Fig. 8. ASIL and DAL target failure rates

Decomposition of ASIL or DAL is possible between two or more independent elements within the system. For example, a system allocated DAL A may be decomposed to two independent aircraft items each assigned DAL B. Hence, this provides the opportunity that a REESS developed for automotive ASIL C applications may be utilised within a decomposed aviation DAL A architecture[7].

The Preliminary System Safety Analysis (PSSA) is a method to evaluate alternative architectures and to select the intended system architecture, derive high-level safety requirements, and to allocate the budgets for hardware failure rates across the system elements. FTA and CCA are utilised; however, it is worth noting that ARP4761 does not suggest conducting a Failure Mode and Effects Analysis (FMEA) during the PSSA activity.

The System Safety Assessment (SSA) is a systematic, comprehensive evaluation of the implemented system to show that relevant safety requirements and target hardware failure rates are achieved.

Both ARP4754A and ARP4761 are classified as guidelines. However, the Federal Aviation Administration (FAA) and European Union Aviation Safety Agency (EASA) recognise ARP4754A and ARP4761 as an acceptable method for compliance with Reg.25.1309 (Federal Aviation Regulations), as specified in Advisory Circular (AC) 25.1309-1[8]. In brief, Reg.25.1309 requires "The equipment, systems, and installations, must be designed to ensure that they perform their intended functions under any foreseeable operating condition"[9].

Reg.25.1309 applies to large aircraft, with corresponding Reg.23.1309 for small aircraft. AC 23.1309-1E[10] describes means for compliance for small aircraft, which again refers to ARP4761 and ARP4754.

As required for automotive, general high-quality processes are also to be followed for aviation, such as those specified by ISO 9001 and AS9100 (Quality Systems - Aerospace - Model for Quality Assurance in Design, Development, Production, Installation and Servicing).

---

[7] Subject to comprehensive analysis and confirmation that the process, methods, and work product expectations of DO-178, DO-254, ARP4754, ARP4761 etc. are achieved

[8] A brief overview is available at https://en.wikipedia.org/wiki/AC_25.1309-1

[9] Full details are available at https://www.law.cornell.edu/cfr/text/14/25.1309

[10] Refer to http://www.faa.gov/documentLibrary/media/Advisory_Circular/AC_23_1309-1E.pdf

2.2 Technical documents

Various technical documents, comprising regulations and guidelines, are published for different product domains and technologies. These documents may be specific to regions (US, EU, APAC, etc.) and countries (China, India, etc.). The primary documents associated with REESS within EU and US regions are reviewed in the following sections.

2.2.1 Automotive domain

UN/ECE Regulation no. 100 'Uniform provisions concerning the approval of vehicles with regard to specific requirements for the electric power train', and specifically Revision 2 Part II of this regulation, 'Requirements of a Rechargeable Energy Storage System with regard to its safety' contains functional requirements and test specifications pertaining to:

- Vibration, thermal shock and cycling, mechanical impact, fire resistance
- External short circuit protection
- Overcharge protection
- Over-discharge protection
- Over-temperature protection
- Emission (of gases) protection

Amendment 5 to this regulation, issued July 2021, adds functional requirements and test specifications, including:

- Overcurrent protection
- Low-temperature protection
- Management of gases emitted from REESS
- Thermal propagation

These requirements allocate to both the physical properties of the REESS (e.g. impact resistance) and/or to the battery management systems (e.g. over-discharge protection). To provide context, an example requirement from UN/ECE Regulation no. 100 is shown in figure 9. Test specifications and acceptance criteria are also specified within the document.

6.8.	Over-discharge protection
6.8.1.	The test shall be conducted in accordance with Annex 8H to this Regulation.
6.8.2.	Acceptance criteria
6.8.2.1.	During the test there shall be no evidence of:
	(a) Electrolyte leakage;
	(b) Rupture (applicable to high voltage REESS(s) only);
	(c) Fire;
	(d) Explosion.
	Evidence of electrolyte leakage shall be verified by visual inspection without disassembling any part of the tested-device.
6.8.2.2.	For a high voltage REESS the isolation resistance measured after the test in accordance with Annex 4B to this Regulation shall not be less than 100 Ω/Volt.

Fig. 9. Example requirement of UN/ECE Reg. 100 Part II

Through the application of the processes and methods of ISO 26262, similar safety hazards and safety requirements may also be derived. Therefore, the developers should seek to ensure consistency between the identification of hazards and specification of safety requirements derived from the application of ISO 26262 with those in UN/ECE Regulation no. 100.

2.2.2 Aviation domain

DO-311A 'Minimum operational performance standards for rechargeable lithium batteries and battery systems', provides requirements, testing, and installation guidance pertaining to:

– Battery protective features
– Battery warning features (over temperature, under voltage, over voltage)
– Charging and discharging protection
– Over-discharge protection
– Mitigation of cell failures
– Venting provisions
– Cell balancing
– State of health function
– State of charge function

Whilst DO-311A was not originally intended to apply to electrically propelled aircraft, there is no specific exclusion, hence in the authors opinion, remains applicable. There are additional technical documents pertaining to REESS, however these are more applicable for small or medium size REESS, not for propulsion, and include:

- AC 20-184: Guidance for the installation of lithium batteries on aircraft
- RTCA DO-347: Certification Test Guidance for Small and Medium Sized Rechargeable Lithium Batteries and Battery Systems
- UL 1642: Standard for Lithium Batteries

There is a level of consistency between the requirements and test specifications of DO-311A and UN/ECE Regulation no. 100 Part II. For example:

> UN/ECE Reg. 100: *"The charging shall be continued until the tested-device (automatically) interrupts or limits the charging"*

> DO-311A: *"The battery system shall automatically inhibit charging when any cell is outside the manufacturer specified voltage or temperature limits"*

Clearly, if a REESS is to be developed with the potential application of both aviation and automotive domains, then an aggregation of the requirements and tests within the various documents is suggested.

3 Exemplar development process and work products

As forementioned, this paper studies the system level safety development activities of a REESS. Exemplar process models, consistent with the expectations of ARP4761 and ISO 26262 are provided in figure 10 and figure 11. The process flows are depicted in a manner to highlight the similarities between the respective aviation and automotive activities. For example, step A1 and A2 for aviation is similar in scope to step A for automotive, step B is similar in scope, and so forth.

Figure 10 provides an exemplar process for ARP4761 'System Functional Hazard Analysis' (SFHA) and ISO 26262 'Concept Phase'. The objective of the process is to:

- Define the system and system functions (step A)
- Identify and classify safety related hazards (step B)
- Specify initial safety requirements (step C)
- Verify the analyses and requirements (step D)
- Produce and approve a report (step E)

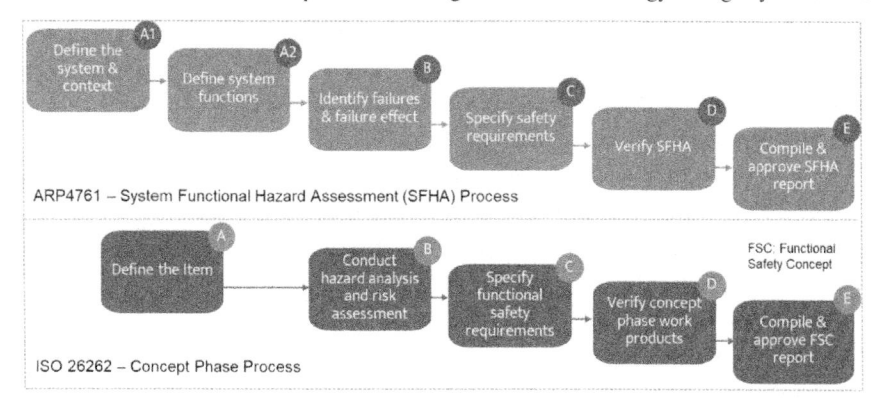

Development of Rechargeable Electrical Energy Storage Systems... 99

Fig. 10. Exemplar process model (Hazard identification and preliminary requirements)

Figure 11 provides an exemplar process for ARP4761 'Preliminary System Safety Assessment' (PSSA) and ISO 26262 'Technical Safety Concept' (TSC). The intent of the process is to:

- Analyse the system for potential faults, failures, and their propagation to hazards (step F)
- Specify detailed safety requirements to prevent, or detect and mitigate, failures (step G). This also includes lifecycle requirements e.g. manufacturing, maintenance and operation.
- Update of the system architectural designs and constrains (step H)
- Verify the analyses, requirements, and design (step I)
- Produce and approve a report (step J)

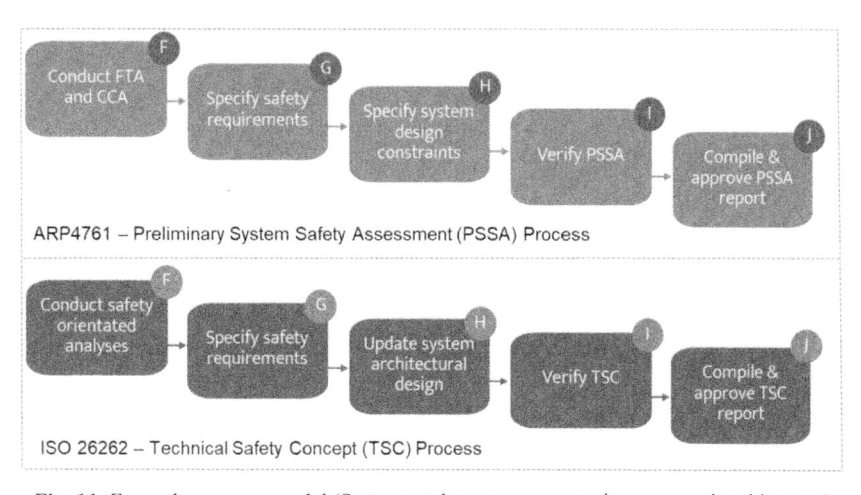

Fig. 11. Exemplar process model (System analyses, system requirements and architecture)

Each of the activities in figure 10 and figure 11 (step A through to step J) will be further described in the following section, including technical examples and comparisons between the automotive and aviation domains.

For this paper we have included specific examples relating to the hazard of 'fire and smoke', caused by thermal runaway (loss of thermal management) of the REESS. This is well defined and understood hazard (Ahrens, 2020) (Ghoshal, 2020) (Sun, Bisschop, et al, 2020). Other hazards, for example 'loss of propulsion', are not discussed within this paper but clearly all hazards must be thoroughly analysed.

3.1 Step A: Define the system and system functions

Objective: To obtain a comprehensive understanding of the system, including regulations, standards, system boundaries, system functions, use cases, and interfaces.

A detailed knowledge of the system of interest must be achieved for the subsequent process steps to be comprehensively conducted. A common approach can be taken, and unified set of work products can be produced for both automotive and aviation domains. Obviously, there is some unique content due to different boundaries, installations, use cases, etc. These points of variation can be documented, so a clear definition of common and unique aspects of the system is achieved.

For REESS, a common preliminary system architecture is illustrated in Fig.. This consists of the battery pack, which is comprised of multiple battery modules, each having a plurality of battery cells and an associated Module Control Unit (MCU); a supervisory Battery Control Unit (BCU); contactors and various sensors including current sensor.

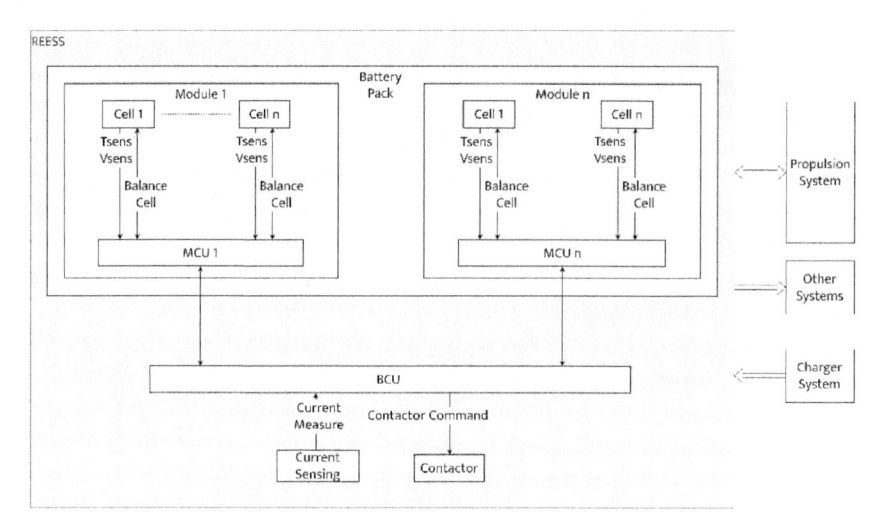

Fig.12. REESS preliminary system architecture

Similarly, a list of common functions can be derived for REESS as shown in figure 13. There may be additional functions which are specific to each domain, these would be derived based on the use case analysis and is outside the scope of this paper.

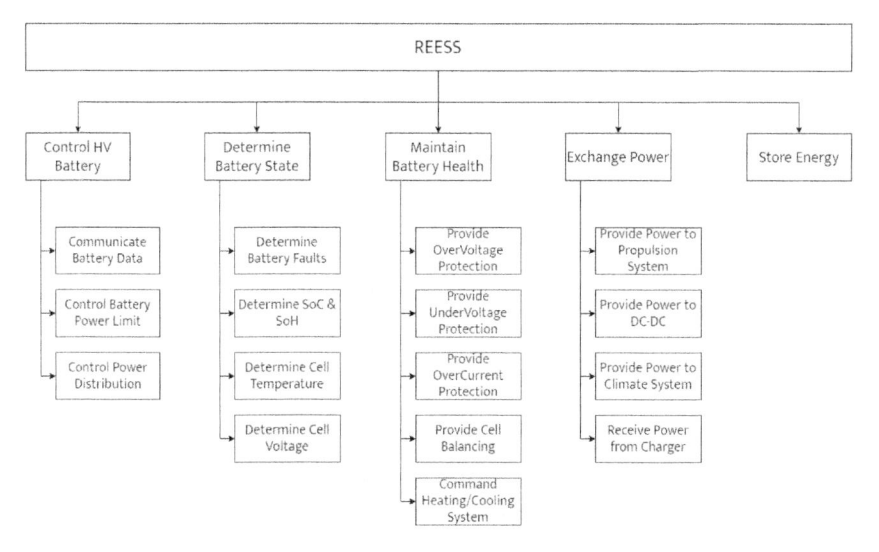

Fig. 13. REESS function tree

Summary: *Step A* can be a common process step and is highly beneficial for identifying and considering the points of communality and variance between the two domains.

3.2 Step B: Identify and classify safety related hazards

Objective: Identify the vehicle and/or aircraft level hazards associated with malfunctions of the system and determine the respective ASIL & DAL classification.

Automotive

A Hazard Analysis and Risk Assessment (HARA) is conducted to identify hazards and evaluate the risk of harm to individuals. The details of the process are well described within ISO 26262, therefore only briefly detailed here to provide context.

The operational scenarios of the intended vehicle application are identified, and consideration for any specific use cases associated with the REESS need to be captured. These include details of manoeuvres (e.g. vehicle speed, cornering), environment (e.g. road gradient, weather), location (e.g. highway, garage), as so forth.

REESS specific scenarios may include plugged-in charging, charging station, and so forth. Examples for REESS are shown in Table 2.

Table 2. Partial vehicle operational scenario list

Scenario ID	Operational Scenario
SN_01	Driving Normal - High speed (>100kph), Acceleration or Cornering
SN_02	Stationary Normal - Vehicle Parked in Public Charging Station, Driver Present, Powertrain OFF, Plugged-in Charging
SN_03	Stationary Normal - Vehicle Parked in Garage, Driver Not Present, Powertrain OFF, Plugged-in Charging
SN_04	Stationary Special - Vehicle Stationary after a Collision with Obstacle, Other Vehicle or Road User. Vehicle approached by fire and rescue team.

The exposure (E) rating is determined for each vehicle operational scenario. The exposure may be rated based on the frequency or duration of vehicle being in that scenario. Examples for REESS are shown in Table 3.

Table 3. Exposure classification of operational scenarios

Scenario ID	Exposure Class	Exposure rating	Exposure Rationale
SN_01	Duration	E4	High speed is a very common scenario on highways/motorways and primary roads.
SN_02	Frequency	E3	A long-distance drive is likely to include some breaks where the driver would plug-in to conduct a rapid charge of the vehicle battery. Driver/occupants may remain inside the vehicle to rest whilst charging.
SN_03	Duration	E2	A few households have garages used to park vehicles. It is unlikely that the driver or others after parking the vehicle would remain inside the garage.
SN_04	Frequency	E1	A collision can happen on a rare occasion.

The severity (S) and controllability (C) ratings are dependent upon the hazard occurring in a particular operational scenario. Severity is rated assuming that there are no safety mechanisms present within the REESS to protect against the hazard. Similarly, the controllability rating depends on how the driver and vehicle occupants would respond in a specific scenario when the hazard occurs. The response of pedestrians and neighboring vehicle occupants are also considered within the analysis.

The examples in Table 4 show the severity and controllability rating for the thermal event hazard caused by a malfunction in REESS within each example operational scenario.

Table 4. Severity and Controllability Ratings

Operational Scenario	Potential Accident (Event sequence)	S	Severity Rationale	C	Controllability Rationale
SN_01	Battery fire occurs. Because of high speed, the driver would take some time to stop the vehicle and evacuate. Panic may also result in road accidents.	S3	If the Li-Ion battery catches fire, severe burns and/or fatalities of the occupants may occur.	C2	Driver may be able to park the vehicle safely, allowing all occupants to evacuate. Other road users would take precautionary measure and maintain distance to avoid harm.
SN_02	Battery fire occurs during public charging.	S3	As above	C1	Occupants likely to smell fumes and exit vehicle.
SN_03	Battery fire occurs. No vehicle occupants. Potential for adjacent buildings / rooms to catch fire.	S3	Severe burns and/or fatalities of household occupants may occur	C2	Fire prevention and detection systems in the property would alert occupants.
SN_04	Impact causes battery to catch fire. Occupants stuck in vehicle due to collision.	S3	Severe burns and/or fatalities of the occupants or rescue service may occur.	C3	First responders will have rescue recovery procedures there may be delays in their arrival.

ASIL is determined for each operational scenario based on the respective S, E, and C ratings (refer to figure 4). The highest ASIL is selected and assigned for the specific hazard. All other potential hazards are evaluated in a similar manner.

In the example of thermal event hazard, the highest rating is evaluated as 'ASIL C' and hence the safety goal inherits this ASIL. The safety goal (Table 5) is the first high level safety requirement on the system (item). There will be safety goals for each hazard, and these are each assigned ASIL A through ASIL D. For this study, the maximum ASIL of all hazards is evaluated as 'ASIL C'.

Table 5. Safety Goals

Hazard Description	Safety Goal ID	Safety Goal	ASIL
Thermal Event with fire	SG_01	Vehicle occupants and others in proximity shall not be exposed to unreasonable risk due to a thermal event caused by battery control system malfunctions.	ASIL C

Aviation

The SFHA is a system level hazard assessment to establish the system safety objectives. The SFHA follows a systematic method of examining the functions and classifying their malfunctions based upon the potential hazard severity.

Flight phases (Table 6) are considered when performing the hazard assessment, since the operational situation of the aircraft will clearly affect the severity of a hazard should a malfunction occur. The flight phases are also critical from the point that it conveys how much time the crew have in hand to maintain safety of flight. Further, the flight time in these phases is documented to aid the quantitative analysis.

Table 6. Example flight phases

FP_ID	Flight Phase	Description
FP_01	Take Off	From the application of take-off power, through rotation, and climb to an altitude above runway elevation.
FP_02	Climb	Any time the aircraft has a positive rate of climb.
FP_03	Cruise	The period following the Climb during which the aircraft is in level flight.
FP_04	Descent	Any time before Touchdown during which the aircraft has a negative rate of climb.
FP_05	Touchdown	From a low height until the aircraft touches down and exits the landing runway.
FP_06	Taxi	When the aircraft is moving on the ground.
FP_07	Standing	Any time before the taxi or after arrival while the aircraft is stationary.
FP_08	Charge	The period when the aircraft is plugged to an external charger.

The prevalent environmental conditions (table 7) also play a vital role in evaluating the risk associated with a hazard, since they may affect flight controllability. These environmental conditions will be considered within the SFHA.

Table 7. Example environmental conditions

EN_ID	Condition	Description
EN_01	Ideal	Dry, warm, high visibility, low wind speed
EN_02	HIRF	A high intensity radiated field (HIRF) of a strength sufficient to adversely affect either a living organism or the performance of a device.
EN_03	Volcanic Ash	Ash cloud
EN_04	Extreme ambient	Less than -10 C or greater than 40 C

The severity of each hazard is systematically determined for combinations of flight phases and environmental conditions. Combinational malfunctions are also considered during this activity. Classification of the hazard is supported by analysing past incidents and experiences, and in consultation with subject matter experts and flight crews.

The example in table 8 shows the effect of loss of the thermal management function of the REESS within an electric aircraft. It can be seen through the analysis that the severity classification reduces for some flight phases as the controllability increases. The environmental conditions are assumed to be ideal for this example.

Table 8. Classification of hazard severity

Failure Description	Flight Phase	Effect	SFHA Classification
Loss of thermal management	Landing Taxi Standing Charging	Loss of thermal management shall lead to overheating of the battery and may eventually result in fire and explosion. Battery venting is provided to release the over-pressure. However, being on land, the fire management services can attend immediately and initiate evacuation procedure.	Hazardous / Severe Major
	Take off Climb Cruise Descent	Loss of thermal management shall lead to overheating of the battery and eventually results in fire and explosion. Battery venting is provided to release the over-pressure. Crew shall proceed for emergency landing procedure. In the meantime, there is a risk of loss of aircraft due to fire hazard.	Catastrophic

Summary: Step B must be a unique process for automotive and aviation, due to the consideration of automotive and aviation operational scenarios; the assessment of controllability by vehicle drivers and their occupants' verse those of highly trained flight crew; and resulting level of harm to persons. There are also some key differences in methods:

- Automotive does not consider combinational failures.
- The derivation of DAL, unlike ASIL, is not decomposed into evaluation of individual criteria (e.g. S, E, C).
- Aviation is not including the need to specify safety goals.

Both automotive and aviation have the hazard of 'thermal event' (fire). The evaluation of risk for automotive is determined as 'ASIL C', and aviation 'DAL A'[11].

---

[11] These ratings are calculated for this study, based upon the vehicle and aircraft definitions and operational scenarios..

3.3 Step C: Specify the initial safety requirement

Objective: Based upon the identified hazards derive the initial (top-level) safety requirements and assign the respect DAL / ASIL classification.

Automotive

The safety goals are decomposed to Functional Safety Requirements (FSR) considering the system architecture design. The FSRs would further specify the fault avoidance or fault detection and mitigation strategy along with transitioning to a safe state.

Fault tree analysis may be a beneficial method to identify the failures which lead to violation of safety goal. In figure 14 the potential failures which lead to the hazard 'Thermal Event with fire' are shown[12]. Constructing the fault tree considering the physical failures which may lead to the hazard assures that the critical properties of REESS are thoroughly understood and documented.

For example, 'failure of over/under voltage protection' (i.e. failure to maintain the battery cell voltage with safe operational limits) may not directly lead immediately to 'fire', but over time damage is accumulated which may eventually lead to fire.

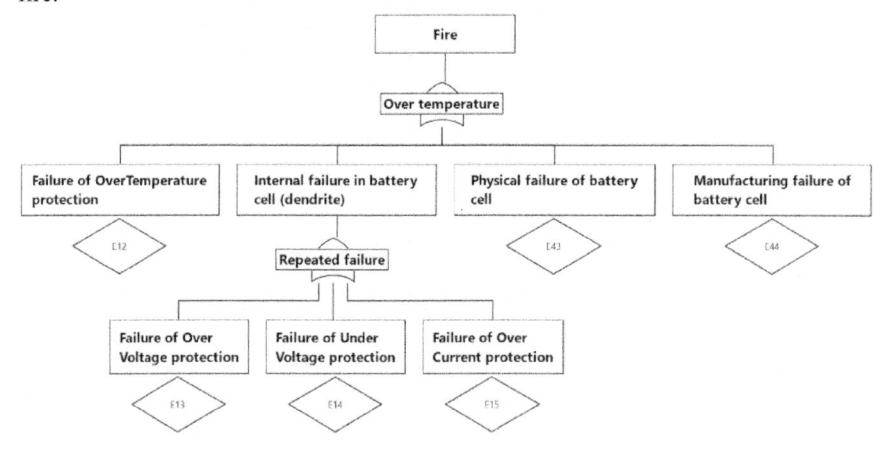

Fig. 14. REESS fault tree for 'fire' hazard[13]

Based upon the fault tree, plus other appropriate methods, the FSR can be derived. Figure 15 shows an example decomposition of the thermal event safety goal into FSR and their respective ASIL rating.

---

[12] This is not an exhaustive list of failures and their failure paths.

[13] The diamond shape represents an undeveloped event. The details of fault(s) which result in the failure(s) is not necessary for this study.

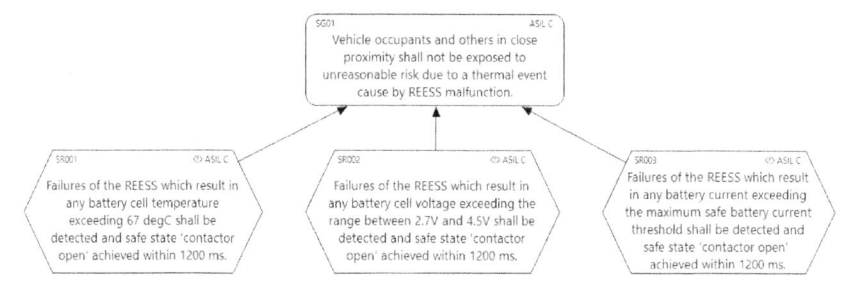

Development of Rechargeable Electrical Energy Storage Systems... 107

Fig. 15. Example safety requirements (automotive)

Aviation

Safety requirements are elicited directly from the SFHA. For each failure condition, safety requirements and target failure rates are directly developed. For loss of thermal management failure, example safety requirements are shown in figure 16.

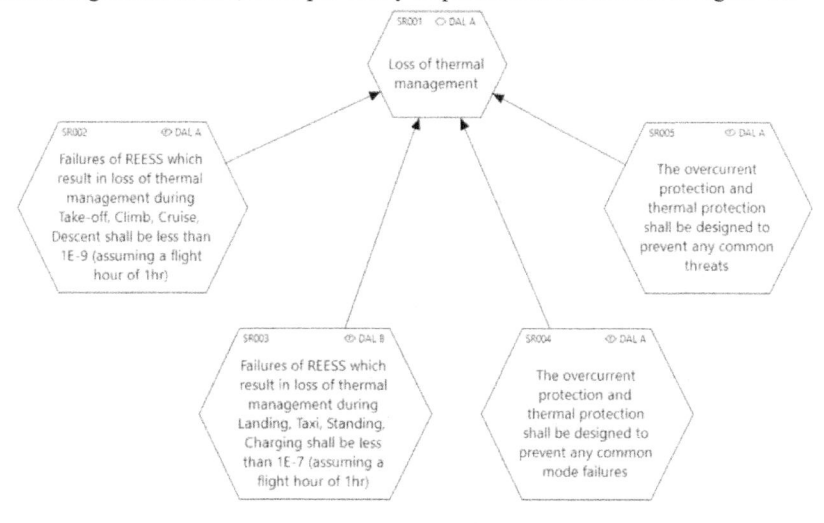

Fig. 16. Example preliminary safety requirements (aviation)

Summary: For aviation, the guideline ARP4761 suggests that the safety requirements are elicited from the SFHA, whereas in automotive, there is an additional level of analyses to be done as per ISO 26262 to systematically derive the functional safety requirements from the safety goal.

The analyses conducted as illustrated in figure 14 helps identification of physical failures internal to the system (e.g. dendrite growth, which is a failure mode of battery where metallic microparticles are collected on the cell electrode when operating in sub-optimal conditions). ISO 26262 also requires specific timing and state transitions to be specified. On the other hand, aviation considers combinational failures within the requirements.

There are best practices in both automotive and aviation which may be utilised to generate a set of safety requirements which are applicable and suitable for both automotive and aviation. It is proposed that a common superset process can be utilised to maximise the potential for reuse and communality across the domains.

3.4 Step D: Verify the analyses and requirements

All work products of Step A to C are verified for quality, technical correctness, and completeness. Verification is conducted by means of inspection (e.g. using a checklist) for both automotive and aviation utilising a largely common set of checkpoints. Example checkpoints are shown in table 9.

Table 9. Example verification checkpoints

Check ID	Checkpoint	Automotive	Aviation
CP_01	All work products are under configuration management	✓	✓
CP_02	The requirements are atomic	✓	✓
CP_03	Each requirement has a unique non-changing identifier	✓	✓
CP_04	The requirements are unambiguous	✓	✓
CP_05	Each safety requirement is assigned an ASIL	✓	
CP_06	Each safety requirement is assigned a DAL		✓
CP_07	The requirements are feasible/achievable	✓	✓

In some instances, simulation or prototyping may be necessary to verify that the analyses and requirements are technically correct, feasible, and achievable.

3.5 Step E: Produce and approve a report

The outcomes Step A to D are summarised in a report. Individual reports are generated for automotive and aviation, but from a common source of information.

3.6 Step F: *Analyse the system for potential faults, failures, and their propagation to hazards*

Objective: Conduct a detailed analysis of the E/E architecture and functions to identify safety concepts which assure achievement of the initial / functional safety requirements.

Automotive

Both inductive (e.g. FMEA) and deductive (e.g. FTA) analyses methods are performed on the functions and system architecture to understand the effect of malfunctions and to design safety mechanisms which satisfy the FSR. Further, Dependent Failure Analyses (DFA) is performed to identify potential cascading and common cause failures within the system architecture.

An example REESS FTA is shown in figure 17. Cut-set analysis identifies potential common cause failures, and in this example the 2nd order cut-set identifies that 'Power supply failure' may be a common cause failure that would affect both over-temperature protection and battery cooling system, leading to hazard of fire.

This will be mitigated or prevented through a design change. Whatever solution is decided, it will then become part of the safety mechanism and subsequently would flow as requirements on Hardware and Software.

The following coupling classes should be considered during the DFA:

- Shared resources (e.g. common power supply)
- Shared information input (e.g. information from a single sensor)
- Insufficient environmental immunity (e.g. heat propagation between elements)
- Systematic coupling (e.g. common design faults)
- Components of identical type (e.g. multiple sensors with same part number)
- Communication (e.g. transmission of information via a shared Controller Area Network (CAN) interface)
- Unintended interface (e.g. incompatible configurations between elements)

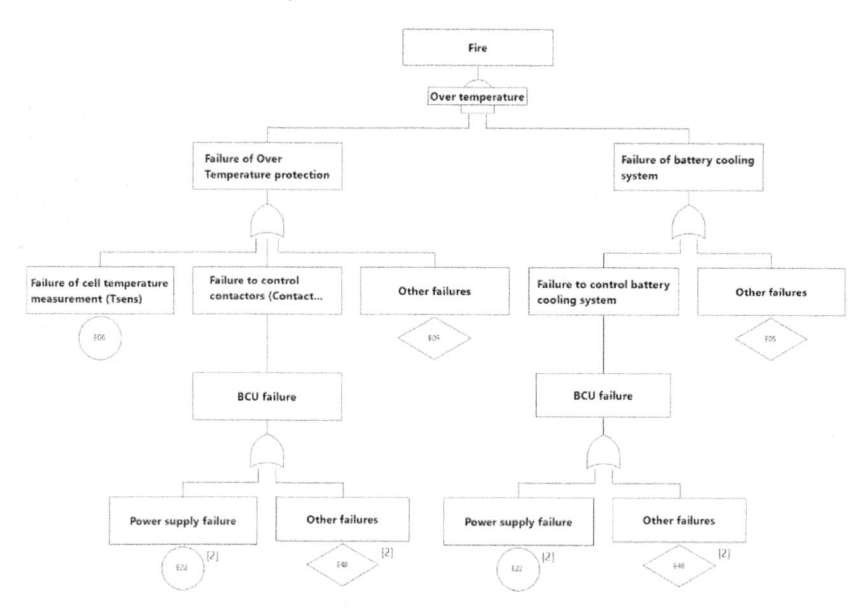

Fig. 17. Fault tree for E/E elements (automotive)

This activity, combined with steps G and H, are iteratively conducted until a final system architecture is derived, including all safety mechanisms and technical safety requirements.

Aviation

The requirements derived from SFHA are used as inputs in PSSA. The intention is to derive a full set of safety requirements on all relevant system architectural elements and to determine if the proposed system design can satisfy the preliminary safety requirements.

For each high-level failure (e.g. loss of thermal management), a deductive analysis (e.g. FTA) is used to identify the system elements which may lead to the high-level failure. Like automotive, FTA is also used to identify common cause failures. The coupling classes of automotive are applicable and supplemented by:

– Zonal Safety Analyses (e.g. elements in same aircraft zones, interactions with other systems in the same zone)
– Particular Risk Analysis (e.g. bird strike)

Additionally, the analysis also supports determination of the failure rate distribution (budgets) within the architecture. One of the factors to include in the fault tree is normalisation of flight phases.

An example fault tree is shown in figure 18 which is derived from an iterated architecture.

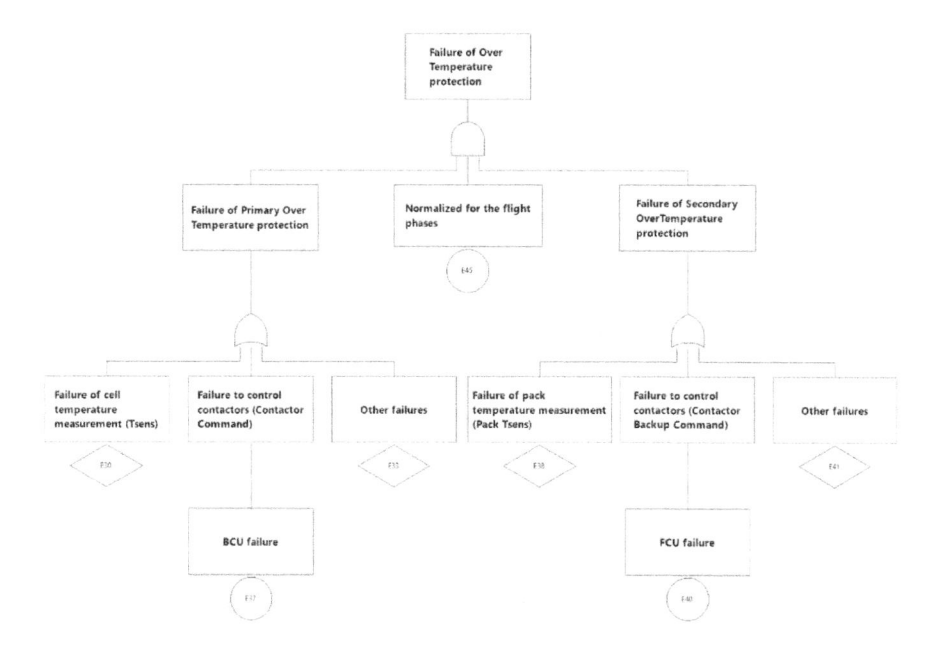

Fig.18. Fault tree for E/E elements (aviation)

Summary: ARP4754A does not suggest FMEA during this phase of development. It is proposed that a common process can be utilised, including FMEA, which will support systematically deriving a set of largely common work products and requirements for both automotive and aviation.

3.7 Step G & H: Specify detailed safety requirements & Update of the system architectural designs

Objective: Specify the safety requirements, revise the system architectural design, and allocate requirement to system elements

Based upon the outcomes of the safety analyses, safety requirements are specified, and architectural designs are updated. Fig. illustrates a potential revised architecture for aviation. Compare this to the preliminary architecture shown in Fig. which remains unchanged for automotive. Additional elements are included for aviation which support achievement of DAL A, through decomposition of DAL A to two independent DAL B elements. A Redundant Monitoring Unit (RMU) is introduced, which is a DAL B item performing independent monitoring of the battery module temperature and independent control of a secondary contactor device.

Alternate architectural solutions exist, for example, an existing Electronic Control Unit (ECU) such as Flight Control ECU may be configured to perform these independent monitoring checks.

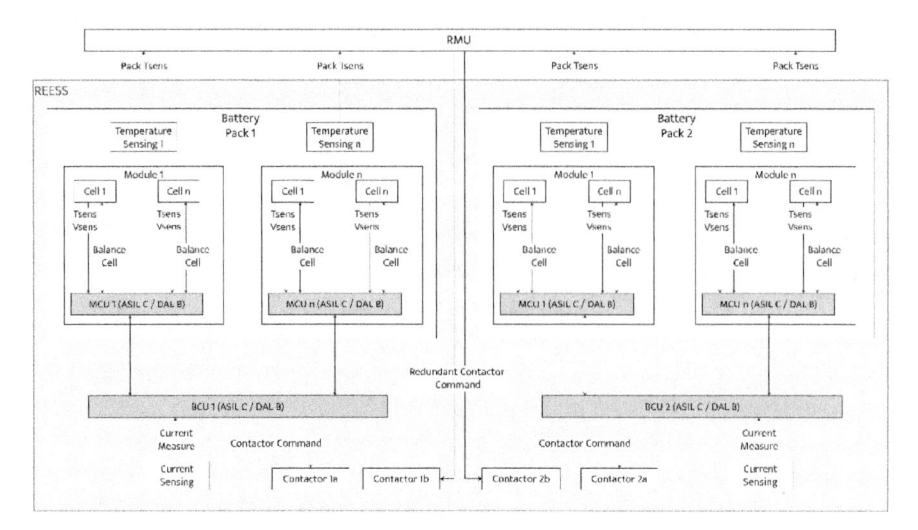

Fig.19. Revised system architecture (aviation)

Based upon the architecture, common requirements can be derived for both automotive and aviation. Equally there are requirements unique to each application.

A small set of example requirements are shown in table 10.

Table 10. Example technical safety requirements

ID	Name	Requirement	Allocations	ASIL	DAL
SR01	Hardware failures metrics	Hardware failures which lead to thermal event in REESS shall achieve the Probabilistic Metric for random Hardware Failures (PMHF) $\leq 10-7$ h-1	BCU	C	B
SR02	Primary safety requirement	The BCU shall detect the battery cell temperature exceeding 67 degC	BCU	C	B
SR03	Redundancy requirement	The RMU shall detect the battery cell temperature exceeding 67 degC	RMU	None	B
SR04	Freedom of interference requirement	There shall be freedom of interference between BCU and RMU	System	None	A

The traceability of these requirements to the high-level safety requirements is represented in figure 20. The intent is to illustrate some example common requirements (SR01 & SR02) and unique requirements (SR03 and SR04) and how these satisfy the original high-level requirements identified in Step B.

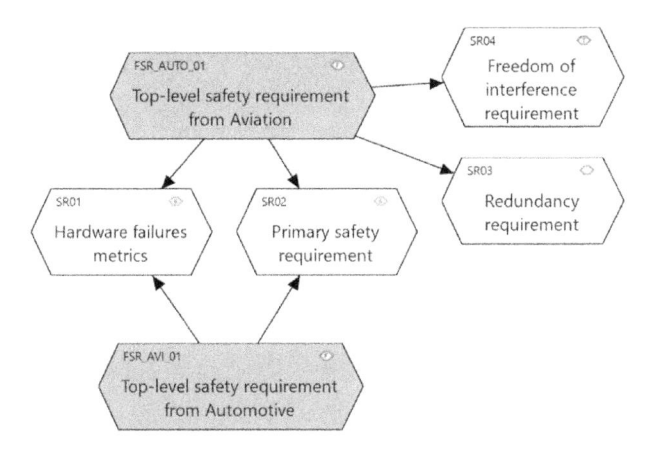

Fig. 20. Requirement traceability (automotive and aviation)

Summary: The decomposition of DAL A to two DAL B elements has enabled the reapplication of an automotive 'ASIL C' REESS within an aviation domain. An independent RMU is introduced, along with other additional elements, to achieve DAL A.

A set of requirements and system architectural designs can be managed across the two domains, clearly indicating the points of commonality and variance.

3.9 Step: I & J - Verify the analyses, requirements, and design. Produce and approve a report

Similar to step D, all work products of Step F to H are verified for quality, technical correctness, and completeness by means of inspection, simulation, or prototyping. The outcomes are summarised in a report. Individual reports are generated for automotive and aviation, but from a common source of information.

Verification, in the form of testing, will be conducted in subsequent process activities. These verify that the system design and requirements are correctly implemented by the system elements. Testing is outside the scope of this paper.

4 Conclusions

This paper has outlined and compared the standards, regulations, and other documents, applicable to development of REESS for automotive and aviation domains. This included both process and technical related documents.

An exemplar process model was then introduced which was used to develop a system design (i.e. the safety requirements and architecture) for a REESS, comparing throughout aspects which can be common, and aspects which need to remain unique for each domain. Opportunities to improve both the aviation and automotive development practices are also identified through sharing of best practice.

For an existing automotive system design to be utilised within aviation the study has shown that while some unique new work products will be required, many of the existing work products may be reused with review and revision to the new context to ensure compliance with aviation regulations.

It must be emphasised that this paper has focussed on ISO 26262, ARP4754, and ARP4761. Comprehensive analysis and confirmation of achievement with the process, methods, and work product expectations of DO-178C, DO-254, DO-330, etc. is clearly necessary.

It is foreseeable that the outcomes of this study may also be applicable to other 'electrification' systems, for example, the charging system and propulsion system.

Acknowledgments The authors wish to acknowledge and thank Dr Ross McMurran and Ms Gayatri Kolhe, 3S Knowledge (3SK) Limited, for their detailed review of the paper, and 3SK for providing the time and resources to undertake this study

Disclaimers This paper is a precis of the outcomes of a study. The numerous examples are included to provide context and should not be taken as technically correct nor complete. All warranties or liabilities, whether expressed or implied, by operation of law or otherwise, are hereby excluded in relation to this paper. This study is independent of any 3SK client project and utilises publicly available and concurred technical content.

References

Ahrens, M. (2020). Retrieved from https://www.nfpa.org//-/media/Files/News-and-Research/Fire-statistics-and-reports/US-Fire-Problem/osvehiclefires.pdf

Ghoshal, A. (2020). Retrieved from https://adreesh-ghoshal.medium.com/how-lithium-ion-batteries-in-evs-catch-fire-9d166c5b3af1

ISO. (2018). ISO 26262-1:2018 Road vehicles Functional safety Part 1: Vocabulary. Retrieved from https://www.iso.org/standard/68383.html

Sun, P., Bisschop, R., Niu, H., & Xinyan, H. (2020). Retrieved from https://www.researchgate.net/publication/338542510_A_Review_of_Battery_Fires_in_Electric_Vehicles

Ethics and Safety for Connected and Automated Vehicles

Paula Palade[1]

Jaguar Land Rover & University of Bradford[2]

Extended Abstract

In September 2020, the European Commission published a set of twenty recommendations (Bonnefon et al., 2020) composed by an expert group (of which the author was a member) on the ethical development and deployment of connected and automated vehicles (CAVs). The report includes twenty recommendations addressing the ethics of connected and automated vehicles, covering dilemma situations, the creation of a culture of responsibility, and the promotion of data, algorithm and AI literacy through public participation. Some of the topics covered are road safety, privacy, fairness, explainability and responsibility.

The 20 recommendations are presented below (Bonnefon et al., 2020):

1. Ensure that CAVs reduce physical harm to persons

To prove that CAVs achieve the anticipated road safety improvements, it will be vital to establish an objective baseline and coherent metrics of road safety that enable a fair assessment of CAVs' performance relative to non- CAVs and thereby publicly demonstrate CAVs' societal benefit. This should be accompanied by new methods for continuously monitoring CAV safety and for improving their safety performance where possible.

2. Prevent unsafe use by inherently safe design

In line with the idea of a human-centric AI, the user perspective should be put centre-stage in the design of CAVs. It is vital that the design of interfaces and user experiences in CAVs takes account of known patterns of use by CAV users, including deliberate or inadvertent misuse, as well as tendencies toward inattention, fatigue and cognitive over/under-load.

---

[1] https://orcid.org/0000-0001-5425-8435
[2] Jaguar Land Rover, Abbey Road, Whitley, Coventry CV3 4LF, UK & University of Bradford, UK

© Paula Palade 2022.
Published by the Safety-Critical Systems Club. All Rights Reserved

3. Define clear standards for responsible open road testing

In line with the principles of non-maleficence, dignity and justice, the life of road users should not be put in danger in the process of experimenting with new technologies. New facilities and stepwise testing methods should be devised to promote innovation without putting road users' safety at risk.

4. Consider revision of traffic rules to promote safety of CAVs and investigate exceptions to non-compliance with existing rules by CAVs

Traffic rules are a means to road safety, not an end in themselves. Accordingly, the introduction of CAVs requires a careful consideration of the circumstances under which: (a) traffic rules should be changed; (b) CAVs should be allowed to not comply with a traffic rule; or (c) CAVs should hand over control so that a human can make the decision to not comply with a traffic rule.

5. Redress inequalities in vulnerability among road users

In line with the principle of justice, in order to address current and historic inequalities of road safety, CAVs may be required to behave differently around some categories of road users, e.g. pedestrians or cyclists, so as to grant them the same level of protection as other road users. CAVs should, among other things, adapt their behaviour around vulnerable road users instead of expecting these users to adapt to the (new) dangers of the road.

6. Manage dilemmas by principles of risk distribution and shared ethical principles

While it may be impossible to regulate the exact behaviour of CAVs in unavoidable crash situations, CAV behaviour may be considered ethical in these situations provided it emerges organically from a continuous statistical distribution of risk by the CAV in the pursuit of improved road safety and equality between categories of road users.

7. Safeguard informational privacy and informed consent

CAV operations presuppose the collection and processing of great volumes and varied combinations of static and dynamic data relating to the vehicle, its users, and the surrounding environments. New policies, research, and industry practices are needed to safeguard the moral and legal right to informational privacy in the context of CAVs.

8. Enable user choice, seek informed consent options and develop related best practice industry standards

There should be more nuanced and alternative approaches to consent-based user agreements for CAV services. The formulation of such alternative approaches should: (a) go beyond "take-it-or-leave-it" models of consent, to include agile and continuous consent options; (b) leverage competition and consumer protection law to enable consumer choice; and (c) develop industry standards that offer high protection without relying solely on consent.

9. Develop measures to foster protection of individuals at group level

CAVs can collect data about multiple individuals at the same time. Policymakers, with assistance from researchers, should develop legal guidelines that protect individuals' rights at group levels (e.g driver, pedestrian, passenger or other drivers' rights) and should outline strategies to resolve possible conflicts between data subjects that have claims over the same data (e.g. location data, computer vision data), or disputes between data subjects, data controllers and other parties (e.g. insurance companies).

10. Develop transparency strategies to inform users and pedestrians about data collection and associated rights

CAVs move through and/or near public and private spaces where non-consensual monitoring and the collection of traffic-related data and its later use for research, development or other measures can occur. Consequently, meaningful transparency strategies are needed to inform road users and pedestrians of data collection in a CAV operating area that may, directly or indirectly, pose risks to their privacy.

11. Prevent discriminatory differential service provision

CAVs should be designed and operated in ways that neither discriminate against individuals or groups of users, nor create or reinforce large-scale social inequalities among users. They should also be designed in a way that takes proactive measures for promoting inclusivity.

12. Audit CAV algorithms

Investments in developing algorithmic auditing tools and resources specifically adapted to and targeting the detection of unwanted consequences of algorithmic system designs and operations of CAVS are recommended. This will include development of CAV specific means and methods of field experiments, tests and evaluations, the results of which should be used for formulating longer-term best practices and standards for CAV design, operation and use, and for directly counteracting any existing or emerging ethically and/or legally unwanted applications.

13. Identify and protect CAV relevant high-value datasets as public and open infrastructural resources

Particularly useful and valuable data for CAV design, operation and use, such as geographical data, orthographic data, satellite data, weather data, and data on crash or near-crash situations should be identified and kept free and open, insofar as they can be likened to infrastructural resources that support free innovation, competition and fair market conditions in CAV related sectors.

14. Reduce opacity in algorithmic decisions

User-centred methods and interfaces for the explainability of AI-based forms of CAV decision-making should be developed. The methods and vocabulary used to explain the functioning of CAV technology should be transparent and cognitively accessible, the capabilities and purposes of CAV systems should be openly communicated, and the outcomes should be traceable.

15. Promote data, algorithmic, AI literacy and public participation

Individuals and the general public need to be adequately informed and equipped with the necessary tools to exercise their rights, such as the right to privacy, and to actively and independently scrutinise, question, refrain from using, or negotiate CAV modes of use and services.

16. Identify the obligations of different agents involved in CAVs

Given the large and complex network of human individuals and organisations involved in their creation, deployment and use, it may sometimes become unclear who is responsible for ensuring that CAVs and their users comply with ethical and legal norms and standards. To address this problem every person and organisation should know who is required to do what and how. This can be done by creating a shared map of different actors' obligations towards the ethical design, deployment and use of CAVs.

17. Promote a culture of responsibility with respect to the obligations associated with CAVs

Knowing your obligations does not amount to being able and willing to discharge them. Similar to what happened, for instance, in aviation in relation to the creation of a culture of safety or in the medical profession in relation to the creation of a culture of care, a new culture of responsibility should be fostered in relation to the design and use of CAVs.

18. Ensure accountability for the behaviour of CAVs (duty to explain)

"Accountability" is here defined as a specific form of responsibility arising from the obligation to explain something that has happened and one's role in that happening. A fair system of accountability requires that: (a) formal and informal fora and mechanisms of accountability are created with respect to CAVs; (b) different actors are sufficiently aware of and able to discharge their duty to justify the operation of the system to the relevant fora; (c) and the system of which CAVs are a part is not too complex, opaque, or unpredictable.

19. Promote a fair system for the attribution of moral and legal culpability for the behaviour of CAVs

The development of fair criteria for culpability attribution is key to reasonable moral and social practices of blame and punishment - e.g. social pressure or public shaming on the agents responsible for avoidable collisions involving CAVs – as well as fair and effective mechanisms of attribution of legal liability for crashes involving CAVs. In line with the principles of fairness and responsibility, we should prevent both impunity for avoidable harm and scapegoating.

20. Create fair and effective mechanisms for granting compensation to victims of crashes or other accidents involving CAVs

Clear and fair legal rules for assigning liability in the event that something goes wrong with CAVs should be created. This could include the creation of new insurance systems. These rules should balance the need for corrective justice, i.e. giving fair compensation to victims, with the desire to encourage innovation. They should also ensure a fair distribution of the costs of compensation. These systems of legal liability may sometimes work in the absence of culpability attributions (e.g. through "no fault" liability schemes).

With respect to road safety, the report suggests 'a minimal requirement for manufacturers and deployers is to ensure that CAVs decrease, or at least do not increase the amount of physical harm incurred by users of CAVs or other road users that are in interaction with CAVs, compared to the harm that is inflicted on these groups by an appropriately calculated benchmark based on conventional driving' (Recommendation 1, Bonnefon et al., 2020) and suggested that the introduction of CAVs requires careful consideration of the circumstances in which they might be permitted not to comply with all applicable traffic rules (Recommendation 4, Bonnefon et al., 2020).

The fact that CAVs should reduce physical harm from road transport is frequently cited as an anticipated benefit of CAVs. In a recent paper, Reed, Leiman, Palade, Martens, & Kester (Reed et al., 2021) tackled this issue suggesting that CAV behaviour should be directed using 'ethical goal functions' (EGFs). This concept embeds the safety and ethical interests of society into functions against which CAV developers can optimise the performance of their systems; developing and assessing these behaviours in simulation and then in real deployments.

References

Bonnefon, J-F., Černý, D., Danaher, J., Devillier, N., Johansson, V., Kovacikova, T., Martens, M., Mladenovic, M.N., Palade, P., Reed, N., Santoni De Sio, F., Tsinorema, S., Wachter, S., Zawieska, K. (2020) Ethics of Connected and Automated Vehicles Recommendations on road safety, privacy, fairness, explainability and responsibility. European Commission, doi:10.2777/035239.

Reed, N., Leiman, T., Palade, P., Martens, M., & Kester, L. (2021). Ethics of automated vehicles: breaking traffic rules for road safety. Ethics and Information Technology, 1-13.

Towards a robust safety assurance process for maritime autonomous surface ships

Gerasimos Theotokatos, Victor Bolbot, Evangelos Boulougouris, Dracos Vassalos

University of Strathclyde[1]

Extended Abstract *The maritime industry has been transformed by the introduction of new technologies pertinent to the 'Industry 4.0' revolution and moves towards the introduction of the Maritime Autonomous Surface Ships (MASS). Research and industrial projects on MASS include the autonomous Yara Birkeland ship design and construction [1], MUNIN [2], AAWA [3], SISU [4], SVAN [4], AEGIS [5], RECOTUG [6] and AUTOSHIP [7].*

The AUTOSHIP project aims at demonstrating the autonomous technology capabilities, thus, pushing the available technology and autonomy levels further on larger size vessels. The Maritime Autonomous Surface Ship (MASS) are classified as a system of Cyber-Physical Systems (CPSs). The introduction of MASS is associated with several challenges related to safety, security and cybersecurity. The safety challenges are attributed to increased complexity related to the unknown interactions in the MASS systems as well as between MASS and environment [8]. Furthermore, cybersecurity has been an important issue, as a cyber-attack can exploit vulnerabilities in the communication links and directly affect the integrity or availability of the data and control systems, leading to accidents [8, 9]. A number of incidences with unauthorised people gaining remote access to the ship control systems has been already reported [10]. All these challenges may jeopardise the introduction of MASSs. Accidents or serious incidents may lead to significant inadvertent coverage by media, backslash from the local community or unbearable litigation costs.

In this study, a safety assurance framework is proposed to support the design of safe, secure and cybersecure MASSs. This framework consists of three phases associated to the three major design phases: preliminary design, detailed design and verification and validation activities. The framework is aligned to the existing guidance for assurance of MASSs and novel technology in the maritime industry,

---

[1] Maritime Safety Research Centre, University of Strathclyde, Glasgow, UK
Correspondence: gerasimos.theotokatos@strath.ac.uk

© Gerasimos Theotokatos 2022.
Published by the Safety-Critical Systems Club. All Rights Reserved.

whereas it also demonstrates sufficient alignment to the existing standards in other industries that can be used for the MASSs design.

Several novel safety, security and cybersecurity analysis methods, fitting into the developed safety assurance framework are proposed and their applicability is demonstrated for theoretical use cases of autonomous ships. The developed framework and the novel methods can be applied in conjunction with other established methods, guidelines and standards. Due to the introduction and use of innovative Key Enabling Technologies (KETs), the proposed framework and methods can be further enhanced and improved by further enhancing existing or developing new methods, especially the ones related to the verification of KETs.

Keywords: Marine autonomous surface ships; Safety assurance framework; Safety methods

References

[1] Yara. Yara Birkeland press kit. 2018.

[2] MUNIN. Maritime Unmanned Navigation through Intelligence in Networks. 2016.

[3] AAWA. AAWA project introduces the project's first commercial ship operators. 2016.

[4] Daffey K. Technology Progression of Maritime Autonomous Surface Ships. 2018.

[5] AEGIS. What if marine automation can take waterborne transport to the next level? 2021.

[6] RECOTUG. SVITZER, KONGSBERG Maritime and ABS join forces to develop the world's first commercial tug to be fully remotely controlled. 2021.

[7] AUTOSHIP. Autonomous Shipping Initiative for European Waters. 2019.

[8] Bolbot V, Theotokatos G, Bujorianu LM, Boulougouris E, Vassalos D. Vulnerabilities and safety assurance methods in Cyber-Physical Systems: A comprehensive review. Reliability Engineering & System Safety. 2019; 182:179–93.

[9] Eloranta S, Whitehead A. Safety aspects of autonomous ships. In: Gl DNV, editor. 6th International Maritime Conference. Germany, Hamburg. 2016. p. 168–75.

A Step-by-Step Methodology for Applying Service Assurance

James Catmur
  J C and Associates

Mike Parsons
  AAIP

Mike Sleath
  Independent Consultant

Abstract *A practical stepwise methodology is outlined based on the SCSC Service Assurance Guidance V2. This is explained using a real example of a highways service (based on a motorway in the UK, the M3); it also includes the services provided by vehicles and drivers when using the motorway. The contracts and agreements are indicated where available, the services identified and then the levels of service assurance flowed down, together with safety requirements. Wrappers (supplementary assurance needs) are identified and elaborated. The advantages of the service assurance approach are then discussed for this very real example and compared to more traditional systems approaches.*

© James Catmur, Mike Parsons, Mike Sleath 2022.
Published by the Safety-Critical Systems Club. All Rights Reserved.

1 Introduction

Many current safety systems rely on functionality provided by services which are designed, developed, operated, and maintained outside the immediate boundaries of the system[1]. In many cases, overall system design is essentially about managing the interactions between various service functionalities, which co-operate to enact a given operational scenario.

This approach is highly applicable to the idea of a highways-service, i.e., the provision of a suitable surface for road vehicles to run on, the signage for the driver, and associated maintenance and development of the road network. All the service provision has to be done to the appropriate level of quality and safety, and tangible assurance artefacts are needed to show this is the case. This work looks at how such a services-view of highways, and the vehicles on them, can lead to a useful, and appropriate, assurance framework.

1.1 Rationale

The main reasons why this approach is useful are:

1. Highways are operated and maintained as a service
2. A highway does not produce or modify anything material for its user
3. A service-based approach to assuring safety provides a different, useful, and important perspective
4. A service-based approach to safety includes the impact of organisations, agreements, and contracts. It is the only safety assurance approach to do so.
5. It is recognised that collaborative working of technology, organisations, people, and processes all contribute to safety and need to be part of the picture
6. A service approach supports the concept of time-limited contracts which are appropriate for road construction, operation, and maintenance
7. There is a significant shift to a service-based approach in many areas of technology and commerce, and hence it is worth exploring an established service delivery example

---

[1] The system here is the overall entity; other sub-systems and services contribute to this overall system.

1.2 Services and the Real World

A service-based approach to procuring functionality or capability is used extensively within the commercial world. Almost everything of significance with safety-related impact is procured and managed as a service (i.e., there will be contracts to maintain, support, and evolve a functionality and for a specific time period). The business case for using commercial services is very compelling:

1. System Architects can easily develop new safety-related capabilities using a Service Based Solution (SBS) building on existing contracts, products, and solutions. They can effectively 'mix and match', joining up known elements to produce what they need.
2. Specialist and off-the-shelf services have reached high levels of maturity, especially in areas such as IT service provision. They are:
    a. Highly resilient and highly available with rich and sophisticated functionality;
    b. Cheap, as the economies of scale and standardised commodity service competition mean lower costs than in-house implementation;
    c. Have growing capabilities as service providers continually evolve their services to varied consumer needs.
3. The trend is set to continue and accelerate as these services expand and mature.

1.3 Scope

This work considers only top-level services for a highway which are: (i) reasonably obvious and (ii) have information available in the public domain.

1.4 Limitations of study

Some contract information is commercially sensitive and not available in the public domain. Where necessary we have proposed some of the missing details. This does not detract from the validity of the method.

2 Introduction to Service Assurance

Service Assurance is a different way of producing trust or confidence in something which is not product-based; it is based on information about organisations, contracts and other information relating to the delivery of the service. For this reason, it is a much more appropriate method of gaining confidence in something where there may be no tangible deliverables. Highways (strictly the journey over the highway) is such an example, where we need to have confidence that it can be undertaken safely but nothing material is "handed over" to the person making that journey in a vehicle.

2.1 Definitions

Term	Definition
Operational Level Agreement (OLA)	Defines the interdependent relationships in support of a service-level agreement (SLA). The agreement describes the responsibilities of each internal group toward other groups, including the process and timeframe for delivery of their services. The objective of the OLA is to present a clear, concise and measurable description of the service provider's internal support relationships.
Service-Based Solution (SBS)	An SBS comprises the systems, organisations, processes, and resources to deliver and manage the services through the duration of the contract life. It may consume other services.
Service Catalogue	A Service Catalogue is the commercial document that lists and describes the services offered for consumption. It is constructed by the Service Provider and typically does not give any service implementation details.
Service Consumer	A Service Consumer consumes (i.e., makes use of) one or more Services
Service Contract	A Service Contract is the legal agreement between Service Provider and Service Consumer. Note that the Service Consumer may not be involved in defining the service or the SLAs at the outset; they may be provided, pre-defined and pre-packaged by the Service Provider on a take-it-or-leave-it basis

Term	Definition
Service Definition	The Service Definition describes the services available for consumption which may include technical and/or commercial aspects. It may include deliverables, prices, contact points, availability, ordering, and processes to request Services. This may include a Service Catalogue.
Service Level Agreement (SLA)	An SLA is the agreement between the Service Provider and Service Consumer that defines the level of service (e.g., in terms of availability, performance, and quality) that the Service Consumer will receive. It often has targets for each service described in the Service Catalogue. It usually specifies responsibilities of the Service Provider and Service Consumer and defines the penalties in the event that the specific targets in the SLA are not met.
Service Provider	A Service Provider provides (i.e., offers to consumers) one or more Services.

2.2 What is a Service?

The way that a Service is normally described or defined is different from the specifications and descriptions more commonly used in safety-related systems. An individual Service (sometimes called a *Service Component*) is typically described by a *Service Provider* via an entry in a *Service Catalogue*[2]. The Service Catalogue usually describes the capabilities/functionality offered to a *Service Consumer* without providing much (or indeed, any) implementation detail, in fact it is unusual for the design and implementation of the Service to be visible to the Consumer. Note that this is a commercial document as well as a technical one and may give information such as the hours a service is available, level of support, etc.

Service Level Agreements (SLA) are used to define the level of service being offered, this may include functional and non-functional properties such as capacity, performance, and availability. SLAs often describe (commercial) penalties on the Service Provider for not meeting key elements of the agreements. Typically, penalties are framed in terms of service credits, but may be also monetary.

---

[2] A high-level document (HE, 2021) which suggests what services may be offered by National Highways.

Service Contracts between the Provider and Consumer provide the overriding legal and commercial picture and typically refer to Service Catalogues, Statements of Work (SoW) and SLAs.

The boundary between a Service Consumer and a Service Provider is typically both an organisational and commercial boundary as well as a technical one. A Consumer may not be involved in the specification and development of a Service and instead may select a commodity or standardised Service (i.e., something already widely available). Alternatively, they may be involved in the creation of new, tailored, or bespoke services.

2.3 Service Context and Service-Oriented Architecture

Figure 1 below gives the context of Service Provision and Consumption:

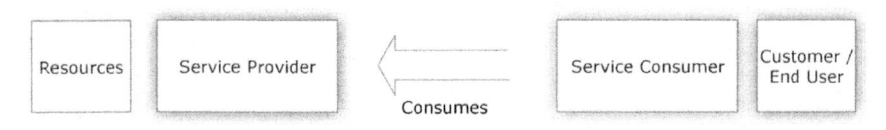

Fig. 1. Context of Service Producers and Consumers

2.4 Simplified View of Highway Services

This paper is concerned with the service analysis of the commercial contracting arrangement with the highway services provider (Highways England, now **National Highways**)[3]. This is necessarily a simplified view but serves as a very useful and real example of how the service assurance approach may be applied.

Note that services are used extensively in a highways network, not only for provision and maintenance of the highway, but also for areas such as breakdown and accident management.

In summary the following are considered:
1. The highway services provider.
2. Other service providers (e.g., breakdown services, accident management) [*not addressed further in this paper other than to acknowledge "Other Services"*].

---

[3] It is acknowledged that most accidents on the road have a significant driver component, and hence we briefly consider the driver and vehicle in section 7.

3. Service consumers (a driver and their vehicle in combination with all other drivers and associated vehicles) [*briefly considered in Section 8*].

3 The Service Assurance Guidance Document v2

The Service Assurance Guidance Document V2 (SAWG, 2021) provides a framework for safety assessment of services, together with principles and objectives for assuring them. The main elements of the document are as follows:

The Introduction explains why services in a safety context are a problem. It covers background aims and scope, and the target audience. The overall approach is that the document is positioned as guidance; it may be used for developing (domain-specific) standards and further guidance for services. It discusses views of what a service is and what service characteristics are. It also introduces service terms used in the document.

The Assurance of Services section begins by introducing some of the challenges of assuring services as a way to describe what is different about services (as opposed to systems) from an assurance view. It introduces further concepts and terms relevant to assurance of services. Finally, it lists some basic assumptions used in the document.

The key part of the document, Service Assurance Principles, states the six Service Assurance Principles, including brief supporting descriptions and explanations. It then defines objectives which support each principle; these are seen as a route of demonstrably meeting the principles. There is also a mapping of the principles to service characteristics.

The concept of Levels of Service Assurance (LSA) is described next. The levels are then used to scope the applicability of objectives, so tailoring what is required for each level of service risk.

The Capturing Justifications and Evidence section provides evidence tables covering aspects of service scoping, design, analysis, implementation, and change. These tables suggest evidence techniques and containers for meeting the objectives. The concept of Assurance Wrappers is introduced and explained. Some further service assurance challenges and some solutions are discussed.

A brief discussion of possible assurance techniques is given in the Analysis Techniques section with the most promising techniques identified for further work.

The document also provides extensive supporting sections including the following topics: (i) Service 'Mode' Changes, (ii) What Happens when Services Go Wrong? (iii) Further work, (iv) a set of Hazop-style guidewords for services, (v) a set of service-related Incidents and Accidents as identified from publicly available sources.

It should be noted that the current guidance document does not provide a workflow; this paper shows one possible step-by-step technique.

4 Methodology

A stepwise methodology has been established on top of the Service Assurance Guidance V2 document[4]. This methodology works well with a Contracting / Service Operator / User model and allows assurance to be explicitly allocated to specific stakeholders. It is also suitable for typical Legal / Commercial models in place for services. The methodology has eight basic steps:

1. Identify Organisations Involved
2. Organisational Provision
3. Determine Service Hierarchy
4. Determine Levels of Service Assurance (LSA)
5. Identify Assurance Wrappers
6. Assurance of the Service-Based Solution
7. Requirements Flow Through Wrappers
8. Meeting the Principles and Objectives

These are explained in more detail below. Section 5 gives a worked example.

Step 1 - Identify Organisations Involved
Initially the organisational hierarchy defined or implied by the contracts that are let to form the delivery of the overall service is identified. Starting with the organisation letting the contract or the prime contractor, the group of organisations delivering the service is expanded and elaborated. Many different organisations may be involved, and most will have some sort of contract or agreement with the prime or next level contractor. It is useful to include organisations directly supporting the service and other stakeholders.

Step 2 - Organisational Provision
The next step in the method then identifies the services being provided by each element[5] of the organisational hierarchy. Often these will be defined by service catalogues or blueprints.

---

[4] This methodology is included in V3, issued in February 2022.

[5] A service-based solution may make use of other things than services (products, subcontracts, etc.)

Step 3 - Determine Service Hierarchy
The third step is to establish the agreements in place at each level of the hierarchy. These are typically contracts which involve a Service-Level-Agreement (SLA). Note that this step should also identify who is accountable and/or responsible for each service within the hierarchy[6].

Step 4 - Determine Levels of Service Assurance
The next step is to identify the levels of service assurance (LSA) across the service hierarchy, i.e., assign an LSA for each element of the service based on the table from the Service Assurance guidance, starting with the top element of the hierarchy. The rule followed is that at least one element below must inherit the LSA of the higher-level element.

Step 5 - Identify Assurance Wrappers
The next step is to identify where there may be assurance shortfalls; this then requires an *assurance wrapper* around the lower-level element, where a wrapper is supplementary assurance that augments that provided by the service provider. A wrapper may be produced by the service provider or a 3rd party, or by the service consumer (road user, in this case) themselves. In this way the wrapper enables a non-safety assured service to be consumed by higher-level service which requires assurance i.e., it "turns a non-safety assured service into a safety-assured service".

Step 6 - Assurance of the Service-Based Solution
The next step is to identify and allocate requirements through the SBS ensuring that safety requirements are met by a service suitable to meet them, i.e., one supported by appropriate assurance information.

Step 7 - Requirements Flow Through Wrappers
This considers how requirements need to be translated. Generally, only non-safety requirements flow through a wrapper, i.e., safety requirements are translated / transformed into non-safety requirements (e.g. from safety to availability). However, it may be that part of a safety requirement can be flowed down to a sub-service provider or supplier, depending on their capability and willingness to accept it.

Step 8 - Meeting the Principles and Objectives
The last step is to go through the objectives contained in the guidance document tables and produce or obtain evidence that shows they are met. Step 8 is mentioned but not elaborated in the worked example in Section 5.

---

[6] It could be presumed that the service provider is responsible but are they also accountable? The service consumer may also have a degree of accountability as they have to consume the service in a responsible manner.

5 M3 Motorway Example

As discussed above there is a need to understand the structure of the services being delivered on the M3 motorway road surface, signs and signals and operation, and how they are organised and assured.

Before doing so, it is useful to understand how road services are managed in England (we use England in this paper, as each UK nation is different). Broadly the roads are divided into main, trunk roads and local, non-trunk roads. The trunk road network outside London is typically managed by National Highways as the Strategic Road Network (SRN). The SRN consists of motorways, dual carriageways, and other major 'A' routes. It should be noted that some motorways and A routes are not part of the SRN and are managed by local highway authorities.

There are several types of motorways with varying age, design, number of lanes and technology. The services provided on each section of motorway will therefore vary. For example, some motorway sections have active signs and signals and are controlled from a control room and patrolled by National Highways Traffic Officers (HETOs[7]), while other sections are not[8].

The M3 motorway is within National Highways' Area 3:

Fig. 2. National Highways Regions[9]

---

[7] Presumably the acronym dates to "Highways England Traffic Officers".

[8] The DMRB (Design Manual for Roads and Bridges) can be found at https://www.standardsforhighways.co.uk/dmrb/

[9] New Civil Engineer, www.newcivilengineer.com

A Step-by-Step Methodology for Applying Service Assurance   133

For this example, we need to understand how the relevant services are delivered within Area 3. Much of the high-level information on these services is within the public domain, although the details of the contracts are not[10]. The high-level services involved in managing the safety of the M3 are:

1. Road Design Service: Major changes to the road are designed to deliver the safety objective [Not developed in this paper]
2. Road Construction Service: The road is constructed to deliver the safety objective [Not developed]
3. Road Monitoring Service: The road is monitored to review its safety performance [Not developed]
4. Road Safety Audit service: carries out audits during design, construction and monitoring [Not developed]
5. Construction Design Management Service (CDM): carries out reviews during design, construction and monitoring [Not developed]
6. **Road Maintenance Service: The road is maintained [developed below]**
7. **Road Operations Service: The road is operated [developed below]**
8. Road Update Service: Any minor changes to the road are subject to a design service [Not developed]

Five of these services shown in a simplified hierarchy in Figure 3:

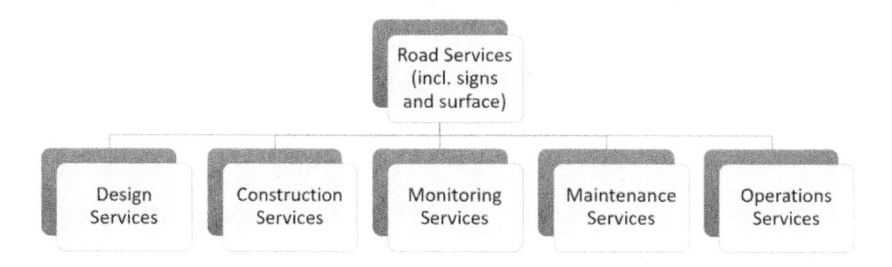

Fig. 3. Principal highway services

To follow the 8-step process, we use two of these services (Maintenance and Operations), and to take the example further we sought information from public sources on the next level of services being delivered.

---

[10] A full list of contracts awarded can be found at: https://www.newcivilengineer.com/latest/50-firms-win-places-on-3-6bn-national-highways-renewals-framework-16-09-2021/

thescsc.org   scsc.uk

Steps 1 and 2

5.1 Road maintenance service

The tender for the maintenance service specified (Area 3, 2020):

> *"Highways England is looking for a suitable contractor to provide and undertake all cyclical and routine maintenance, incident response, defect rectification and severe weather delivery on the all-purpose trunk roads and motorways within Hampshire, Surrey, Berkshire, Oxfordshire, Wiltshire and part of Buckinghamshire, not including the M25, (Highways England Area 3).*
>
> *Duties will also include maintenance of roadside technology, the provision of traffic management for the contractor's maintenance and incident response activities (including traffic management for others), where instructed by Highways England and the maintenance of depots."*

The services are therefore broadly split into the 4 categories shown in Table 1 and Figure 5, below:

Table 1. Four broad categories of road maintenance services

1	Routine maintenance
2	Defect repair
3	Incident response
4	Severe weather response

This broad list is broken down in more detail in (Area 3, 2020):

Table 2. Detailed categories of road maintenance services

1. Highway maintenance work	2. Repair, maintenance and associated services related to roads and other equipment
3. Road traffic-control equipment	4. Miscellaneous repair and maintenance services
5. Tunnel lighting	6. Bridge and tunnel operation services
7. Tunnel linings construction work	8. Tunnel operation services
9. Construction work for highways	10. Engineering Services
11. Roadworks	12. Electrical installation work

| 13. Road-maintenance works | 14. Installation of signalling equipment |
| 15. Surface work for highways | |

The information available publicly shows that the Kier Asset Support Contract involves an alliance that is in place for Area 3 and it gives an idea of next level of companies / service teams involved in delivering the Service:

- National Highways
- Kier
- Aggregate Industries
- Carnell
- Crown Highways
- Tarmac
- WJ, Chevron
- CLM
- Volker Laser
- PDS
- R&W
- Chambers Southern
- Enims
- Roocroft
- R & C Williams
- KJ Thulborn
- Waltet
- Maurer
- Camps Highways LTD
- AC Landscapes & Treeworks.

5.2 Road Operations Service

National Highways operate the M3 through their own services or with the assistance of external services:

- The Regional Control Centre (RCC) service delivers monitoring and operational management of the road (Figure 4, below):

136 James Catmur, Mike Parsons, Mike Sleath

Fig. 4. The control room, RCC

- The National Highways Traffic Officer (HETO) service delivers response to incidents and works with the RCC to manage incidents
- The RCC and HETO services work with other external services (Police, emergency services, recovery services) to support response to and management of incidents

Using this National Highways M3 service structure we can elaborate the service hierarchy as shown below in Figure 5 (note that the Vehicle Separation and Vehicle Control services are brought forward from section 8):

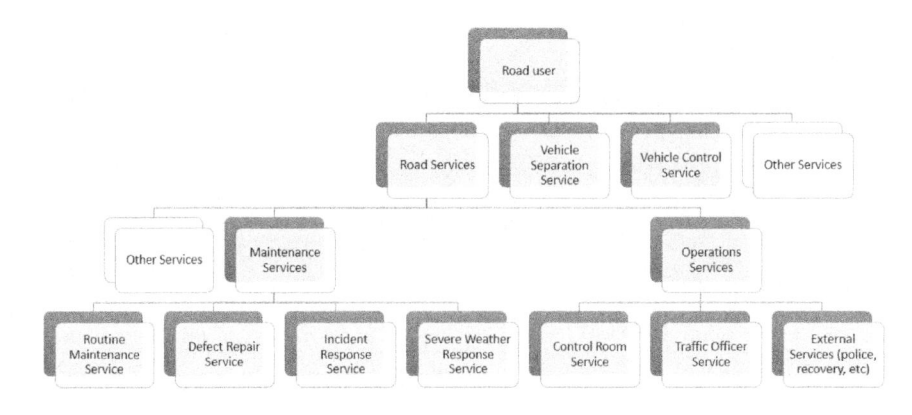

Fig. 5. Next level of highway services

A Step-by-Step Methodology for Applying Service Assurance 137

5.3 Service contracts

Step 3

A service hierarchy will be supported by explicit and implicit service contracts.

The contract between the maintenance service provider Kier to National Highways will contain explicit requirements on matters such as frequency of inspection, maintenance schedules, time-to-repair and time-to-respond. We do not have access to the detail of the contract but for the sake of this example we assume that it is sufficient to deliver the four sub-services identified in Table 1.

The RCC and HETO services are both internal to National Highways and, while again we do not have access to the detail of the internal agreements/contracts/SLAs/OLAs, we assume that they are sufficient to deliver the two services.

There is no publicly available information on any explicit contracts between National Highways and other external bodies (road recovery, police), although there are agreements and implicit contracts in place between them.

There are no contracts in place between the road user and the vehicle control and vehicle separation service providers, although many road users assume there is an implicit one if they pay their taxes and obey the highway code.

Figure 6 shows the service hierarchy along with the agreements that we understand to be in place to support service safety assurance.

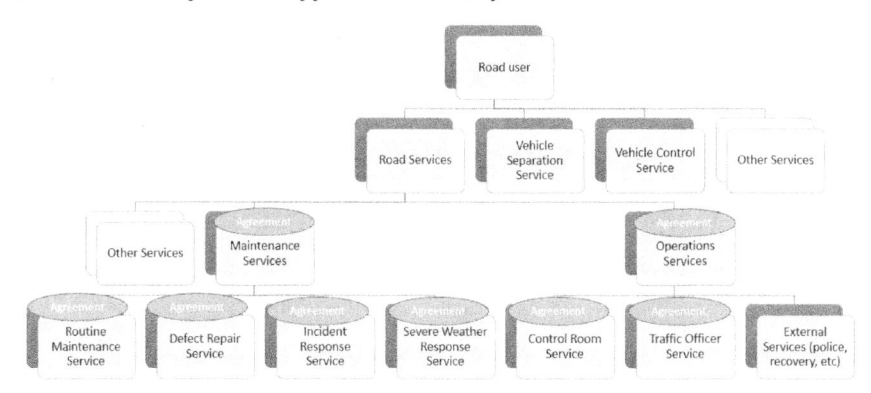

Fig. 6. Service agreements in place

5.4 Levels of Service Assurance

Step 4

Once the service hierarchy has been identified the Levels of Service Assurance (LSA) are mapped onto it. This uses LSA 0 to 4, which relate the LSA to the potential impact of a failure in the service, as shown in Figure 7.

Level of Service Assurance	Definition (Service **Consumer** View)
LSA 0	No safety aspects present in service
LSA 1	Minor safety aspects with little impact of failures (minor injury possible but unlikely)
LSA 2	Safety aspects with some impact of failures (several injuries possible)
LSA 3	Significant safety aspects with service with major impact (could indirectly lead to multiple injuries or a single death)
LSA 4	Service is safety-critical: service failures could have catastrophic impact (could directly lead to multiple deaths)

Fig. 7. Levels of Service Assurance (SAWG, 2021)

We can see that at the top of the service hierarchy the LSA is 4, as road accidents have the potential to cause multiple deaths. This LSA is then delivered by lower-level services, with at least one lower-level service delivering the same LSA as the higher-level service. Figure 8 shows a possible flow down of LSAs to the lower-level services. It should be noted that this is an example, and not a definitive view of the LSA mapping.

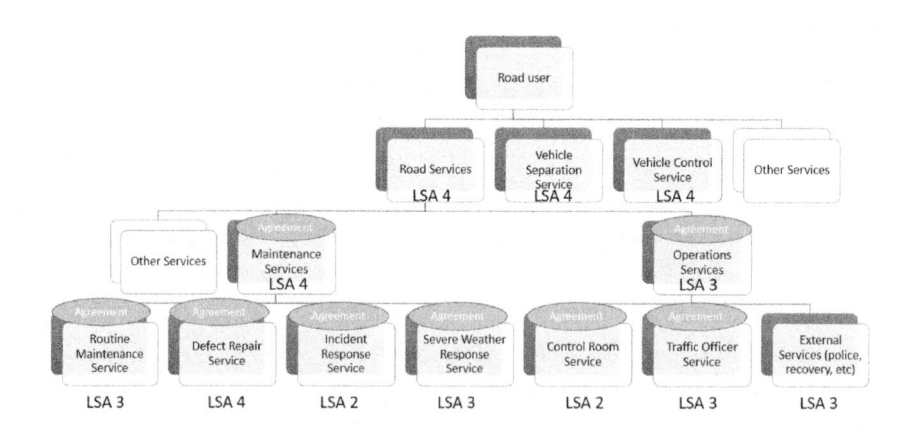

Fig. 8. Possible LSA hierarchy

Figure 8 thus shows an example of how the flow down of the LSA could be satisfied. As an example, the defect repair agreement would provide timescales of the detection of different categories of defects and the expected repair time; allowing for road closures and rapid repair when major defects occur.

5.5 Wrappers

Step 5

Where there is no agreement available to provide such assurance (or the agreement does not provide sufficient assurance) a 'Wrapper' can be used to assist in the service assurance process. The wrapper looks for or creates additional assurance that the service will deliver the required LSA. Figure 9 shows possible service interfaces where a wrapper may be required[11].

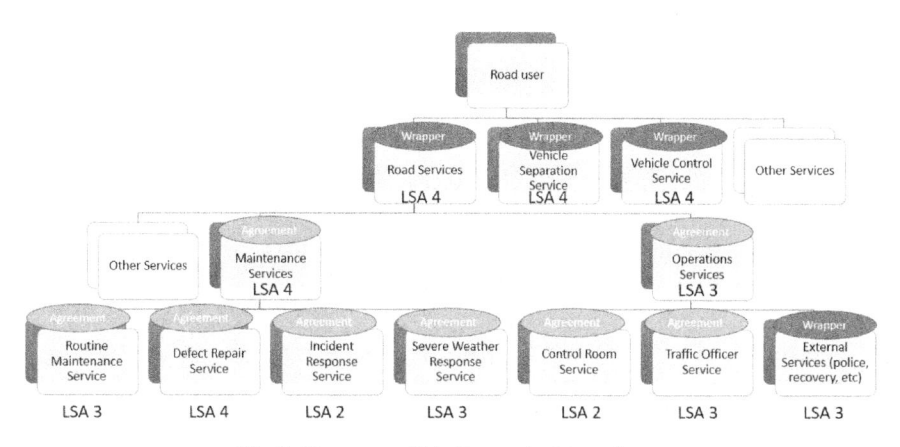

Fig. 9. Wrappers within the service hierarchy

In instances where agreements exist, wrappers may still be required in order to achieve the desired service assurance, through providing evidence that supplements the agreement. For example, an agreement may contain an average response time for a particular activity and the wrapper might supplement this through:

- the maximum response time
- how the performance of the requirement will be monitored over time
- actions required should performance fall below a certain level

As an example, a possible wrapper for the 'vehicle separation service' could provide assurance that:
The driver:
- has passed their test;

---

[11] Note that this proposal does not mean that the agreements provide all the assurance required for the other services, and if done for real, further wrappers may be needed.

- is aware of stopping distances and following distances;
- knows the latest version of the highway code;
- is alert when driving, and
- is not distracted.

The vehicle (where fitted):
- driver assistance systems are active, and
- the vehicle brakes are maintained and working.

5.6 Overall structure of assurance

Steps 6 and 7

Using the example discussed above service assurance can be delivered by showing at each level how the requirements in the agreement deliver the assurance, using a wrapper where required. Using Routine Maintenance Service, it could be shown that the times and actions in the agreement provide assurance that a suitable surface will be delivered, or the road closed for the Defect Repair Service to take action. This may be done by reference back to standards on such matters or established benchmarks for the action.

When this is combined with wrappers the overall assurance can be developed, as shown in Figures 10 and 11. These are two hypothetical examples; in the figures an 'X' indicates a service interface where the agreement and wrapper do not provide sufficient assurance and there may be a need to seek further evidence to demonstrate assurance.

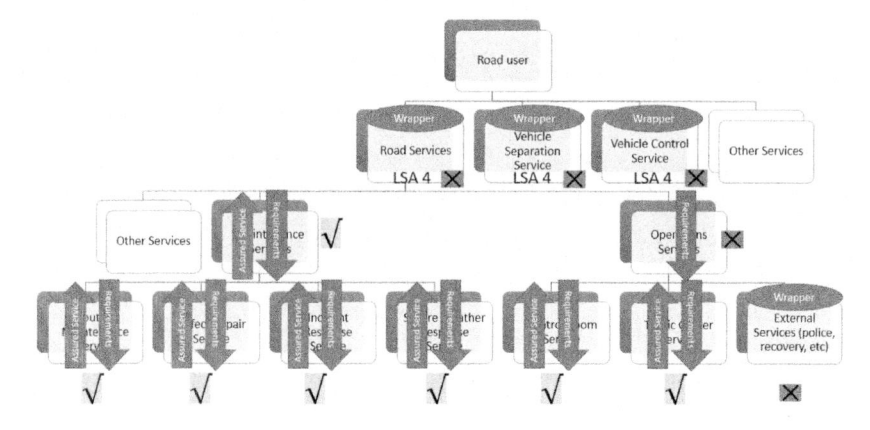

Fig. 10. Example of the flow of service assurance, with hypothetical assurance shortfalls

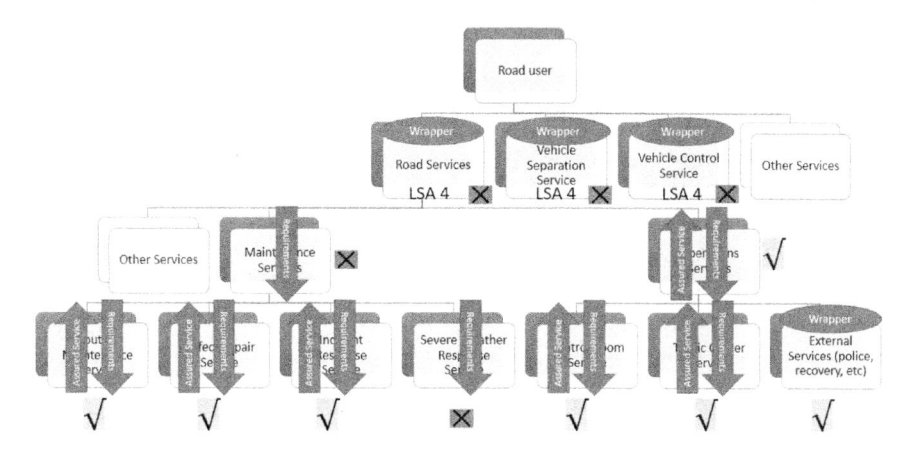

Fig. 11. Further example of the flow of service assurance, with hypothetical assurance shortfalls

5.7 Assurance Evidence

Step 8

Assurance evidence for each service will need to be assembled, with each service developing a logical, service-dependent structure for its evidence. GG104 (DRMB, 2021) requires that each service identifies its safety baseline and safety objective, with a focus on leading indicators when possible. The evidence that the service meets its performance targets may come from operation of the service (for example, mean time to repair defects) or from leading indicators (e.g., operator response time to incidents, or results of user acceptance testing). Projects within National Highways are already developing evidence of service assurance, with a growing focus on leading indicators. The need for such evidence will grow as the focus on service assurance increases, building on the historical focus on compliance with standards and technical assurance. This all supports the evidence required for Step 8 of the process.

6 Evolution of National Highways Approach

National Highways has been developing its safety approach moving from compliance with standards and rules, through assessment of technology to assessment of services. The approach will need to consider the hazards and risks related to both technical systems (for example, the signs and signals) and the services that are required to support them (for example, the signalling rules). This is a strategy that will continue to develop as the highways become even more service-focussed and as vehicles begin to deliver more services (for example, vehicle separation via adaptive cruise control or vehicle control via lane keeping and intelligent speed adaptation.)

Recent examples of the steady move towards more service assurance include the use of STPA (Systems-Theoretic Process Analysis) to evaluate the assurance required in the delivery of fault support services, safety assurance of the change to a new control service, and the requirement for there to be an assessment of the risks related to the delivery of drone-based imaging services.

The National Highways standard on risk assessment, GG104 (DMRB, 2021), includes the assessment of services within its scope (see Figure 12) and there is a need to continue to develop approaches for the assurance of these services. We hope that this paper will show an approach that may be of use to National Highways.

GG 104 Revision 0		1. Scope
1.	**Scope**	
	Aspects covered	
1.1	The approach set out in this document shall be applied to determine the level of complexity of any activity that does or can have an impact on safety risk, either directly or indirectly, for any of the populations on the motorway and all-purpose trunk roads.	
NOTE 1	Activities that do or can have an impact on safety risk for any of the populations on the motorway and all-purpose trunk roads include:	
	1) planning, preparing, designing, constructing, operating, maintaining and disposing of assets (examples of direct, with nothing or no one in between influences on safety risk);	
	2) revising Highways England requirements and advice documents and all procedures, policies and strategies (examples of indirect influences on safety risk).	

Fig. 12. GG104 Scope

7 Conclusions

This section draws some general and specific conclusions from the example looking at the advantages and disadvantages of the approach[12].

---

[12] It is noted that space is limited in this paper and only an outline of the technique can be presented. Also, the example presented above does not yet include details of Step 8 of the process which is to show conformance to the tables in the Guidance Document.

The main advantages of a services-based approach to assurance over more traditional systems focussed approaches are likely to be:

1. It is based on the commercial reality of what is actually in place to provide a safety-related service, i.e., contracts, agreements and delivery organisations within a service hierarchy.
2. It explicitly deals with shortfalls in assurance and proposes to fill the gaps with 'wrappers', i.e., supplementary assurance, or identifies the need for further work.
3. It avoids detailed systems-focussed analysis, as the systems are usually within a service delivery and therefore largely hidden.

There are also some identified disadvantages of the approach:

1. It is as yet a largely untested and untried approach, although it has been applied to some studies
2. Some steps in the process are inter-dependent (for example, Steps 1 and 2) so may need to be linked in the process; considering them as separate steps did help the process
3. It needs sight of commercial information which may not be freely available[13]
4. Assurance of services is a new technique which is not familiar to system safety practitioners
5. Thinking of situations, vehicles, (or indeed, vehicle tyres) in terms of services takes some mental agility which requires practice[14].

8 Further Work

This section outlines further work for this example and highlights ongoing and future work in the service assurance area.

---

[13] Approaches to manage this are beyond this paper but may include abstraction, views of contracts with financial information removed, use of trusted 3rd parties, etc.

[14] In writing this paper the authors had significant debate about this!

144   James Catmur, Mike Parsons, Mike Sleath

8.1 New Guidance Version

It is noted that V3 of Service Assurance Guidance document will be issued at SSS'22 in February 2022, (SAWG, 2022). This version will include further details on service-based safety analyses. It would be useful to revisit this example when the next version is finalised and consider which of the service analyses would be of benefit.

8.2 EU Research Programme

An EU-funded Marie Curie Training Network research project is being proposed in late 2021 on *Trust In Complex Services*. This will provide funding for a set of PhD students working on service assurance topics throughout Europe, including the UK. This work will also reference the Service Assurance Guidance documentation and provide constructive input for continued development of the approaches mentioned.

8.3 Work within the SAWG

Work is underway within the Working Group on a service-assurance approach to the safety assurance of a railway level crossing. This work is progressing, and it is hoped that it can be included in V3 of the guidance.

8.4 Including the Driver and Vehicle

This paper has not examined driver and vehicle aspects in detail, including the interactions with other road users. However, the driver is known to be the biggest cause of accidents, so it may be useful to include this aspect.

So how might a services approach include the driver? The driver normally controls the vehicle to maintain safe distances from other vehicles and infrastructure while progressing on the journey[15]. An approach might be:

1. The *Road Surface Service* is provided by the Highways Provider and offers a suitable road surface on which the vehicle may travel. Strictly the consumer

---

[15] It is noted that most some modern vehicles have a variety of systems to assist the driver in this role.

of this service is the vehicle tyre, as that is the only part which touches the road
2. The *Road Signage and Warnings Service* is also provided by the Highways Provider, giving information to the driver (e.g., via roadside signs or gantry information displays). In this way the driver is the consumer of these services.
3. The driver is shown here to provide the *Vehicle Separation Service* which is consumed by the vehicle (i.e., by the vehicle controls). This service keeps a safe distance between vehicles on the road. It is noted that many modern vehicles also have automatic distance-keeping functions, hence this service would be provided by both the vehicle and the driver.
4. More generally the driver also provides a *Vehicle Control Service* which provides things like throttle and steering control. This is then consumed by the vehicle. Again, it is noted that many modern vehicles are delivering aspects of this service though functions such as lane keeping, intelligent speed adaptation, wrong way driving alert, etc.)

This is illustrated in Figure 13:

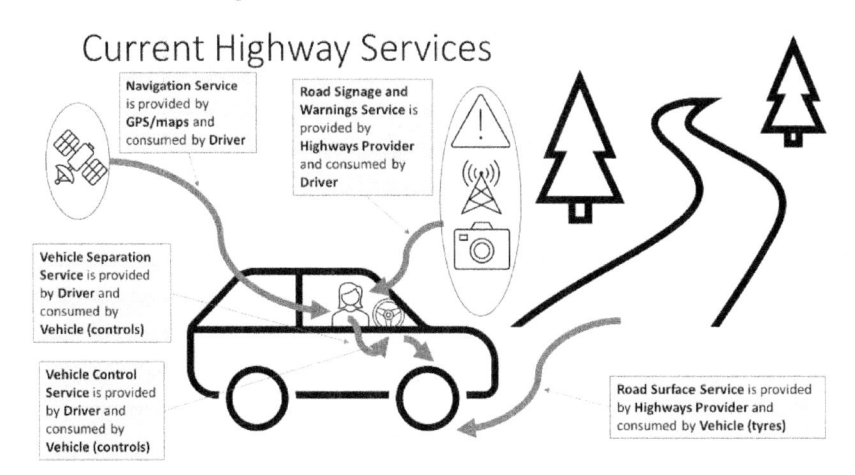

Fig. 13. Simplified View of Highways Services with Vehicle and Driver

It is recognised that as more control functions are placed within the vehicle (ultimately with fully autonomous control) then the driver has less of a role and the services currently provided by the driver transfer to the vehicle[16].

The list below[17] shows the functionality already available in a modern vehicle that, in the hierarchy, are delivered as services.

- ACCQA (Adaptive Cruise Control with Queue Assistance, radar and camera)
- HWA (Highway Assist)
- TJA (Traffic Jam Assist)
- RCTA (Rear Cross Traffic Alert)
- LCA (Lane Change Assist)
- BSD (Blind Spot Detection)
- AHBC (Active High Beam Control)
- LKA (Lane Keeping Aid)
- LDW (Lane Departure Warning)
- TSR (Traffic Sign Recognition)
- DAC (Driver Alertness Control)
- SAP+ (Semi-Automatic Parking)
- Theft and stolen vehicle notification via telematics
- AVSL+ (Automatic Vehicle Speed Limiter)

Hence with increased autonomous functionality, the services might be represented as in Figure 14:

---

[16] However, for now, the question is do we need more assurance of the driver to improve road safety? The answer is probably yes as this could have a big impact on the accident rate. What sort of things could this involve? There could be e.g., additional ad-hoc or regular testing (e.g., re-test every 10 years). There could also be attempts to improve driving culture, noting regional and national behaviours. alcohol testing kits that inhibit start if alcohol detected above set limits, Advanced Driver Tests, etc. There could also be changes to the Highway Code.

[17] From safety and driver assistance features, Lynkco.com, https://www.lynkco.com/en/car/car-specifications-features#1a5ec65a-a17d-4fea-951e-05ed65b07d4d;3

Updated Highway Services

Navigation Service is provided by **GPS/maps** and consumed by **Vehicle**

Road Signage and Warnings Service is provided by **Highways Provider** and consumed by **Vehicle Sensors**

Vehicle Separation Service is provided by **Vehicle** (...) and consumed by **Vehicle (controls)**

Vehicle Control Service is provided by **Vehicle** (planning and routing) and consumed by **Vehicle (controls)**

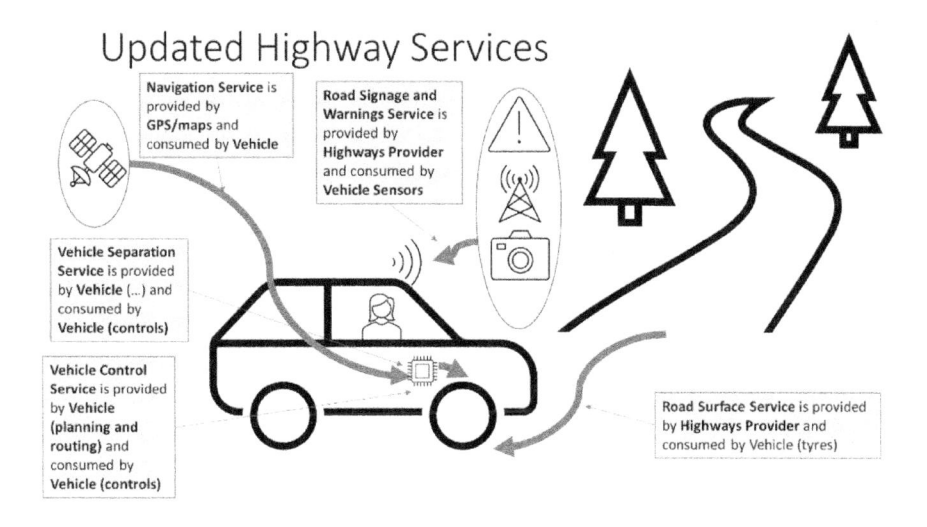

Road Surface Service is provided by **Highways Provider** and consumed by Vehicle (tyres)

Fig. 14. View of Highways Services with Vehicle and Driver with Increased Autonomy

Acknowledgments Thanks to Andy Whitehead and Kevin King for review comments. The work of the SCSC Service Assurance Working Group (SAWG) has been instrumental in developing this methodology.

Disclaimers Views expressed in this paper are those of the authors and not necessarily those of their organisations

References

Area3 (2020), Bidstats, Area 3 Asset Delivery Maintenance and Response (MR) Contract https://bidstats.uk/tenders/2020/W43/737345542, accessed September 2021

DMRB (2021), DMRB: Standards for Highways, GG104, https://www.standardsforhighways.co.uk/dmrb, accessed October 2021

HE (2021), Customer Service Plan, 2021-22 Highways England, https://highwaysengland.co.uk/media/yjbdok5h/customer-service-plan-2021-22.pdf, accessed October 2021

SAWG (2021) Service Assurance Guidance V2 by the SCSC Service Assurance Working Group, SCSC-156A, February 2021, SCSC, https://scsc.uk/scsc-156A

SAWG (2022) Service Assurance Guidance V3 by the SCSC Service Assurance Working Group, SCSC-156B, February 2022, SCSC, https://scsc.uk/scsc-156B

The German and Belgian Floods in July 2021

Peter Bernard Ladkin

Causalis Ingenieurgesellschaft mbH

Bielefeld, Germany

Abstract *Many people died in flooding in Germany and in Belgium in the week of 2021-07-12 to 2021-07-18. The rainfall was foreseen, and warnings generated by the European Flood Awareness System. I relate the events and the environment in detail, consider the sociotechnical systems in place to deal with them, and possible improvements to mitigate the consequences of a repeat rainfall event.*

1 Introduction

In Germany, over 180 people died in the floods in Rhineland-Palatinate (RP) and North Rhine-Westphalia (NRW) states in the week of 2021-07-12 to 2021-07-18 (BPB 2021, Wikipedia-2021-European-Floods 2021, Wikipedia-2021-Hochwasser 2021). The regions hit hardest were two watersheds of the Rivers Ahr and Erft, whose sources are in the northern Eifel massif, a group of low mountains in mid-Western Germany, extending into Belgium and Luxembourg. In Belgium, over 40 people died (Wikipedia-2021-Hochwasser 2021, Belga 2021).

The Eifel, although not high mountains, the highest point being less than 750m above sea level (ASL), can be quite rugged, with steep-sided valleys – an attractive place to visit. The flooding events along the Ahr and along the Erft had different characteristics.

West of the North Eifel (Nordeifel) lies the Hohes Venn (Fr. Haute Fagnes, NL. Hoge Venen, E: High Fen). The Hohes Venn is indeed fenland, with bogs and suchlike. The highest point is the Signale de Botrange, also the highest point of Belgium, at 694 m ASL. The Signale is indicated just under the "e" of "Hohes Venn" in Figure 1[1].

---

[1] There is also a microwave tower, which is used to transmit data between the London and Frankfurt Stock Markets – said to be marginally faster either than satellite or optical fibre cable.

© Peter Bernard Ladkin 2022.

Published by the Safety-Critical Systems Club. All Rights Reserved.

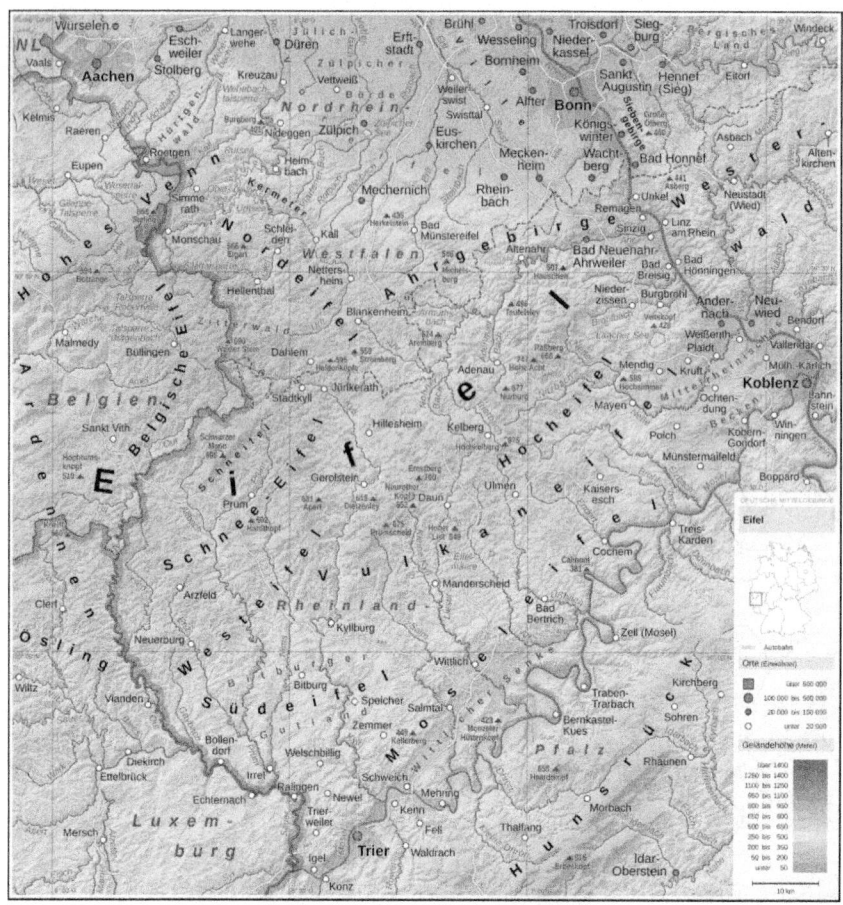

Fig. 1. Topographic Map of the Eifel
Credit: Thomas Römer, CC BY-SA 3.0

North of the Signale de Botrange and west of the German town of Monschau lies the German-speaking Belgian city of Eupen[2]. Eupen is on the Weser/Vesdre river, along which many towns flooded on 2021-07-14. West of Eupen on the Vesdre is Verviers and then Pepinster, Chaudfontaine and the city of Liège, all of which suffered in the floods.

2 Germany

---

[2] German-speakers are recognised in Belgium as the third cultural population after the French-speaking Walloons and the Dutch-speaking Flemish, and have their own Parliament in Eupen.

Germany possibly suffered the most of the countries affected by the intense rainfall. There were over 180 deaths (Wikipedia-2021-Hochwasser 2021) and significant physical damage. I shall consider the events along four watersheds, the Rivers Ahr, Erft and Swist and Rur.

2.1 The River Ahr

The River Ahr rises near the town of Blankenheim, about one-third of the way down the map in Figure 1 and a little to the left of centre. The Ahr is more easily seen on the rainfall map Figure 14, in which it is highlighted. The Ahr is part of the eastern watershed of the Eifel. It starts in the state of NRW, in Blankenheim, and flows ENE into the state of Rheinland-Pfalz, to and through the towns of Altenahr and Bad Neuenahr-Ahrweiler, eventually joining the Rhine between the towns of Remagen and Sinzig. It is almost 90 km long, with a watershed said to be some 900 km^2 (Wikipedia-Ahr, no date). Before this flood, the Ahr was mostly known for its Pinot Noir vineyards and the superb wine they produce.

The designation of the "source" of such a mountain river is often social. Stream/river topography is dentritic near the origins, and to pick one "dendrite" as source and the others as tributaries is mostly a historical/social action. If there is a spring, that will often be designated as "source". How and where a river is named is not necessarily of interest to hydrologists.

Various communities along the Ahr valley, in particular the small town of Schuld, about midway between Blankenheim and Altenahr, were particularly badly hit. Some 100 people lost their lives. Particularly noteworthy structural damage occurred in Altenahr, where the river makes a number of meandering turns around steep terrain[3]. Figure 2 shows Bundesstrasse 267 (B267)[4], the main highway in Altenahr, emerging from a tunnel – and disappearing because it has been washed away. The view is to the NW; the river runs from left to right. The rail bridge to the right looks intact; I don't know whether it is. The second bridge has a section on the south end washed away.

The topography and physiography here play the following kind of role. A large watershed on higher land drains down small valleys into a larger valley with a larger river. All these channels are topographically constrained. The flow resistance along the length of the main river is quite large, and water draining in from the watershed when it reaches the main river has nowhere to go but up.

---

[3] Further pictures of the devastation along the river may be found on the WWW pages of the Bonn newspaper General Anzeiger (GA-Pictures-of-Ahr 2021).

[4] A Bundesstrasse is an intercity road maintained by the German Federal Government, rather than the State in which it is located. It is the equivalent of a British A-category road. The "A" designation in Germany is reserved for Autobahnen, motorways, and is equivalent to the British M-category.

Most notably from a hydrologic point of view is that there is not a lot of "sealed soil" in this watershed – paved-over or built-over land. Where and how high floods occur is a natural phenomenon more than it is a result of human activities such as building and canalisation and biotope modification. The destruction in the towns is, of course, another matter.

Fig. 2. The Remains of Tunnelstrasse (B267) in Altenahr After the Flood
(Credit: picture alliance/ZB/euroluftbild.de/Klaus Göhring)

2.2 The River Erft

The River Erft rises in the town of Holzmülheim, not shown in Figure 1, about 8 km NE of Blankenheim as the crow flies (the 10km scale is bottom right), so not far away from the source of the Ahr. The Erft flows north, through Bad Münstereifel, Figure 3, which suffered problems similar to those of towns along the Ahr, for similar reasons.

The Erft then flows into a flood plain north of the Eifel. The flood plain is large, extending northwards for – well, up to the Dutch coast; east to the Rhine, and West to Düren (see Figure 4). The Erft flows through Euskirchen (bottom middle of Figure 4) at the southern edge of the plain, then through Erftstadt, denoted in Figure 4 by two of its municipal districts, Lechenich and Liblar, bisected by Autobahn A1, and on north-northwestwards, eventually to join the Rhine at Neuss (not shown). Figure 14 shows the run of the Erft more clearly.

Fig. 3. Bad Münstereifel from the Air
Credit: Wolkenkratzer CC BY-SA 4.0

The landscape in Erftstadt is very different from that in Bad Münstereifel, as we can see in Figure 5. It is very flat; largely floodplain. However, Figure 6 shows a large landslide caused by the flood, less than 1 km east of Lechenich in the municipal district of Blessem, between Lechenich and Liblar.

How can a landslide possibly happen here? It's flat, flat, flatland. Land washed away. Where to?

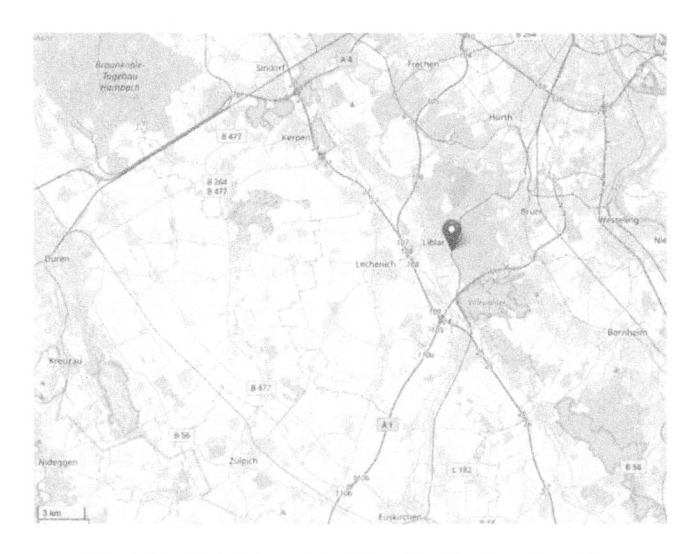

Fig. 4. The Plain West of the Rhine (Credit: OpenStreetMap).

Figure 7 shows Erftstadt in more detail. Between Erftstadt-Lechenich and Erftstadt-Liblar lies Erftstadt-Blessem, where the large landslide took place. The Autobahn A1 (red) and the Erft (blue, just to the east of the A1/A61) are clearly visible.

Fig. 5. Erftstadt-Lechenich (Credit: Howi [Willi Horsch], CC BY 3.0)

Fig. 6. The Landslide in Erftstadt-Blessem, looking South
(Credit:Rhein-Erft-Kreis)

Fig. 7. Details of the Municipality of Erftstadt (Credit: OpenStreetMap)

So how do you get a landslide, as happened in Blessem, on a flat floodplain? Figure 8 gives a good hint. Gravel pits are present, in Figure 8 top middle, clearly marked in the map in Figure 7 as Kieswerk Blessem ("Kies" means gravel in German) just north of Blessem. The sides of a gravel pit are not particularly stable. Too much surface water, as in the flood, can erode the sides and wash earth away and into the pit. There is a (once-)magnificent white building in Figure 8 middle, Burg Blessem, which was recently restored, and is now mostly no more (Burg Blessem was a "Wasserschloß", a fortified structure surrounded by a moat, which in this case had been filled in.)

Various other streams or rivers flow through Erftstadt. Lechenich was also quite badly damaged, presumably through flooding on the Rotbach ("red stream"), which flows through the centre. Again, how so? There is no gravel pit here. This is a flood plain, so surely water can just spread out? Actually, not. There is a fair amount of sealed soil in the city districts of course, and streams and rivers are canalised, including the Erft as well as the Rotbach.

The watershed here is not topographically constrained, but constrained through human modification, through canalisation. The results are similar: with significant resistance to flow generated by downstream water mass, there is nowhere for the water to go but up. When that is in the middle of town, damage results.

Fig. 8. Erftstadt-Blessem with Burg Blessem, Looking North
(Credit: Rhein-Erft-Kreis)

2.3 The Swist

Between the Erft and the Rhine lies the Swist, sometimes called the Swistbach ("Bach" means stream or creek; there is not usually a large flow). It rises in Kalenborn, in the Eifel about 3 km N of Altenahr, and flows N, parallel to the Erft, joining the Erft just before Bliesheim, near the junction of the Autobahns 61 and 553 some 2-3 km S of Erftstadt. Again, Figure 14 shows the run clearly. The Swist is just over 43 km long.

One of the left tributaries of the Swist, the Steinbach, undergoes a couple of name changes before joining the Swist just after the small town of Miel. The Steinbach was dammed in its upper watershed by an earth dam, the Steinbachtalsperre ("Tal" is valley, "Sperre" is dam). I say "was" for good reason.

On 2021-07-14 around 20:00 the dam was overrun. The overrun gouged deep into the earthworks (Figures 9 and 10). There is a timeline of this event on Wikipedia (Wikipedia-Steinbachtalsperre 2021, Section "Hochwasser 2021"). Luckily, the dam was not breached.

A local construction company owner, Hubert Schilles, heard by telephone call that the outflow was obstructed, and drove to the dam in his 30-tonne excavator to free it up, as the lake was spilling over the top (Stegmann 2021). For good reason, Herr Schilles is now considered locally a hero. The lake has since been drained and the remains of the earth dam removed.

Various towns along and around the Swist were flooded[5]. The town of Heimerzheim was particularly badly hit. The titles on the photos from the General Anzeiger, which also double as brief commentary, say the water was 2.5 m high in the Hauptstrasse ("Main Street"). That suggests, by comparison with other photos of the Swist in its channel, that there was at least 4 m of high water. It has been claimed that no warning was given to the residents (Kalischek 2021).

Fig. 9. The Dam on the Steinbach Showing Extensive Damage
(Credit: picture alliance/dpa/David Young)

The Steinbachtalsperre lies in the Euskirchen District. Heimerzheim lies in a different district, namely in the west of the Rhine-Sieg District ("Rhein-Sieg-Kreis"), which surrounds the former West German capital city of Bonn, which lies on the Rhine to the east. The rainfall in most places in the Rhine-Sieg District was of the order of 50 mm in 24 hours. The lake behind the Steinbachtalsperre overran the dam late on 2021-07-14, which is why flooding was worse along the Swist than in other places in the Rhine-Sieg District. The lake is fully pictured in the General Anzeiger (Westbrock 2020). It is 17 m deep, according to the article.

---

[5] Many good pictures of the destruction along the Swisttal, and some of the destruction in Bad Münstereifel, are to be found with the Bonn newspaper General Anzeiger (GA-Pictures-of-Swisttal 2021)

Fig. 10. The Steinbachtalsperre, Showing a Pedestrian Railing for Scale
(Credit: picture alliance/Geisler-Fotopress/Christoph Hardt)

The Swist is usually a stream, but there is a history of flooding. Floods in 1961 led to the watershed being engineered, and the stream/river canalised. There were further floods in 1984, and some isolated flooding in 2009 (Wikipedia-Swist, no date).

Heimerzheim lies in the municipality of Swisttal ("Gemeinde Swistal") of the Rhine-Sieg District. The municipality has its own WWW site (Swisttal, no date) and a municipal council ("Rat") (Swisttalrat, no date). Two questions arise. First, why little or no warning was given to locals, if that is indeed the case? Second, what systems were in place for monitoring for and reacting to a breach of the upstream dam?

Some of the politicians in the council live in the places they represent[6]. When they say there was no warning, we can presume communication was somehow broken at the district-to-municipality link, or even at the state-to-district link. A particular coordination issue arises in Swisttal with a spillover at the Steinbachtalsperre, because the lake and dam lie in the Euskirchen District, whereas the Swisttal council, the Gemeinde Swisttal, is part of the Rhein-Sieg District. But according to Wikipedia (Wikipedia-Steinbachtalsperre 2021), the story is as follows.

---

[6] I tried contacting the responsible people in the Gemeinde Swisttal to get more information, to no avail.

At 16:35 on 2021-07-14 the lake was full. At 17:00 the Bezirksregierung Köln (the Regional Government in Cologne, which has by law overall responsibility for the management of all the NRW watersheds coming into the Rhine S of Cologne) was informed. At 18:10 the Controller at the Mission Control Centre of the Disaster Protection Agency informed the Landkreis Euskirchen (Euskirchen District government) and the Bezirksregierung Köln of the impending, unavoidable overrun of the dam. At 18:42 the Mayor of the Gemeinde Swisttal ordered the sirens along the Swisttal to be activated, via the Mission Control in Siegburg (responsible for the Siegburg-Bonn District, on whose western edge Swisttal sits). She also chairs the water management agency responsible for the Swist and upstream, the WES-Wasserverband.

At 20:00 the dam was overrun. At 21:00 evacuations began downstream in the Euskirchen district, utilising also boats and helicopters; 4,500 people. Two hours later, evacuations began in the Rhein-Sieg District (which includes Gemeinde Swisttal) further downstream; 7,700 people, including some in Heimerzheim.

We can see that there are many different agencies involved in this disaster warning and relief function: the regional government; the Disaster Protection Agency; the Control Centres of two different Districts; the water management agency; and above all local politicians and the district and municipality level. The state governments and district agencies in Germany are legally responsible for public safety, so these politicians are, indeed must be, a mixture of politician and civil servant in order to execute such required functions. Whether this distribution of responsibility, not only between various agencies but also amongst politicians and civil servants who are not necessarily formally trained in disaster relief, is a help or a hindrance is not so clear at the time of writing. Analysis of the functioning of the disaster warning and relief activities pursuant to these events will be complex.

During the floods on the Somerset levels in England some years ago, as well as the floods in York and the Pennines watershed in general, the perils of canalisation were well discussed in Britain. They are related to the perils of "sealed soil", paved-over ground such as roads which water can only flow along, not sink in, as it can in (some) farmland or forested soil. One could make similar observations about canalisation and sealed soil here. (But that, of course, was not the main problem. The main problem was the extreme rainfall.)

2.4 Two Overflows on the Rur

The Rur river drains the Hohes Venn, the high point of the Eifel, east and then runs north, highlighted in Figure 14 (Wikipedia-Rur, no date). The Rur runs through Monschau, into the Rursee with its dam (Rurtalsperre), which may be seen left middle in Figure 1, then north out of the Eifel to Düren and then NNW through Jülich to join the Maas/Meuse at Roermond ("Roer" is the Dutch name for the Rur). Roermond is roughly west of Düsseldorf. The Rur is not to be confused with the more well-known Ruhr, which is a right tributary of the Rhine.

There was a problem on the Rur in the flatlands shortly before Roermond, where a dam in the town of Wassenberg breached (Tagesschau 2021). Wassenburg is shown in Figure 11, middle. The dammed part of the river may be seen in two lakes just west of the "W" in the name "Wassenberg".

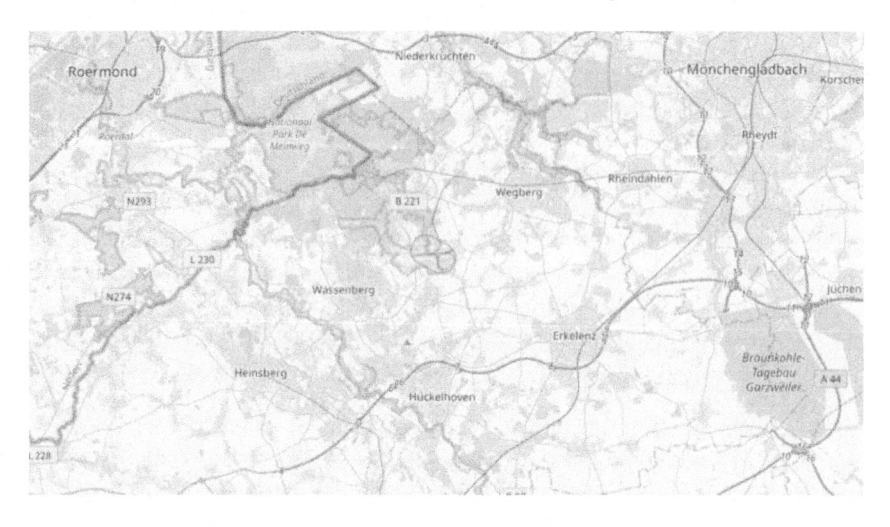

Fig. 11. Wassenberg on the Rur (Credit: OpenStreetMap)

Also, on 2021-07-16, two days after the inundation, the dam on the Rursee ran over (the Rursee is in Figure 1, east of Eupen and Roetgen) threatening flooding in Düren and Jülich, whose vulnerable areas had been evacuated, but it didn't come to that (Knaack 2021).

3 Rainfall

The forecast at 2021-07-14 at 00:00 UTC and the actual rainfall on 2021-07-14 is shown in the left, resp. right frames of Figure 12 (Magnusson et al 2021).

Even the 6-hour forecast did not predict the intensity of the rain on the Eifel (at 50mm-75mm per m^2) whereas it seems to have been more like 100mm-150mm per m^2 (Figure 14). It correctly forecast the extreme rain further east, in Hagen, where a particular station in the Nahmer valley recorded over 200mm (see below).

The German and Belgian Floods in July 2021   161

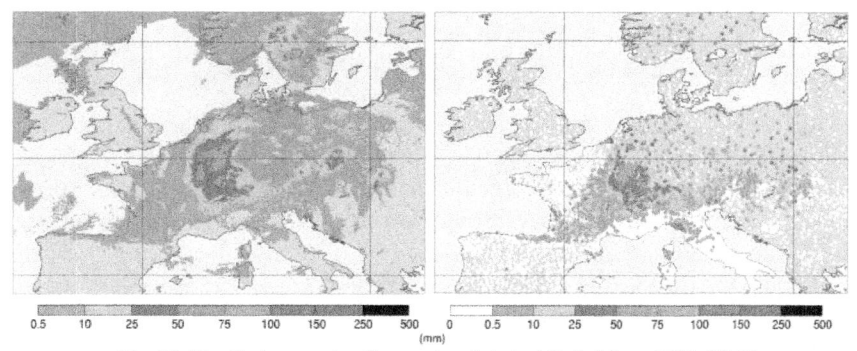

Fig. 12. The Early-morning Forecast and Actual Rainfall on 2021-07-14
(Credit: Magnusson et al 2021, ECMWF 2021)

The ERA5 reanalysis[7] of the actual rainfall from the Copernicus Climate Change Service (C3S) (Magnusson et al, 2021) is shown in Figure 13. Left is the rainfall in block form for the period 2021-07-13 at 0600 UTC to 2021-07-15 at 0600 UTC. Right is the collection of rainfall events from January 1950 to August 2021, showing the number of events (vertical axis) exhibiting given amounts (horizontal axis). The red bar denotes the rainfall in the period 2021-07-13 at 0600 UTC to 2021-07-15 at 0600 UTC. The orange bar denotes the rainfall for the same length period starting one day later: 2021-07-14 at 0600 UTC to 2021-07-16 at 0600 UTC. Notice that the red bar exhibits almost twice the largest amount of rainfall otherwise occurring in over 71 years.

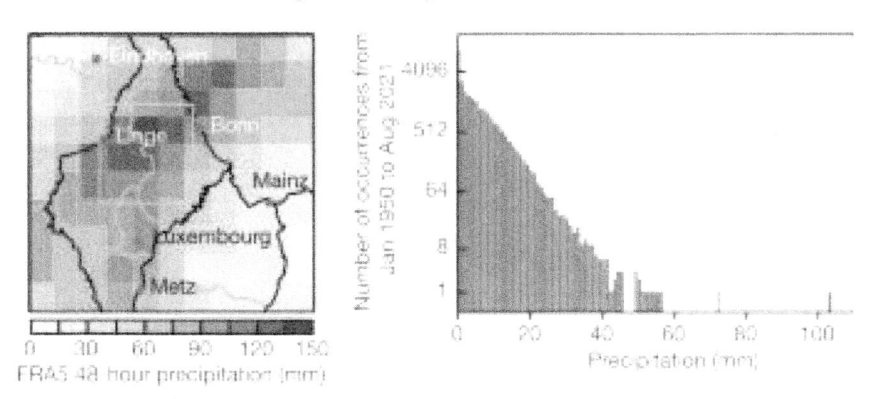

Fig. 13. The ERA Rainfall Reanalysis and Comparison With Previous Events
(Credit: Magnusson et al 2021)

---

[7] "ERA5 is the fifth generation ECMWF atmospheric reanalysis of the global climate covering the period from January 1950 to present. ERA5 is produced by the Copernicus Climate Change Service (C3S) at ECMWF." (ECMWF, no date)

3.1 Rainfall on the Eifel, North and East

Figure 14 shows rainfall amounts in the 24 hours from early on 2021-07-14 to early on 2021-07-15, from the German Weather Service recording stations. Most of the central Eifel, the Vulkaneifel District, shows over 100 mm/m^2. Near the source of the Ahr at Blankenheim, 129 mm/m^2 was measured. East of that, a measurement 150 mm/m^2 is recorded. 1mm of water in a square metre is exactly 1 litre, so mm/m^2 and lt/m^2 (also written l/m^2) are identical measures. Note that rainfall in the area of the Rhine, on the right of Figure 14, are much lower, 26 mm/m^2 in Andernach to 56 mm/m^2 in Bad Honnef, with the notable exception of 154 mm/m^2 in Köln-Stammheim.

The rainfall immediately to the north of the Eifel, in Euskirchen and Rhein-Erft Districts (including Erftstadt), in NRW, was heavy but more moderate, around 114 mm/m^2 in the flatlands north of Euskirchen and then much less further north (with the exception, noted above, of Köln-Stammheim, whose 154 mm/m^2 I think is the highest recorded by the German Weather Service during the event).

Kachelmannwetter.com reports there was actually rainfall of 145 mm/m^2 inside 12 hours in Köln-Stammheim (Ruhnau 2021). Reifferscheid, near Adenau in the Eifel (Figure 14, middle, about 1/3 of the way up), is reported in Wikipedia to have recorded 207 mm/m^2 in 9 hours, but I have not been able to source this otherwise (Wikipedia-2021-European-Floods 2021).

The city of Hagen lies some 60 km NW of Cologne, on the Western end of the Sauerland, a low-mountainous region which is the source of the Ruhr river, which flows through a pass just north of the city, on its way W to Witten, Bochum, Essen, and Duisburg, where it flows into the Rhine, 55 km N of Cologne. According to Radio Hagen at 10:36 on 2021-07-14, the station[8] on the Nahmer stream (Nahmerbach) had recorded 211 mm/m^2 of rain in the previous evening and through the early morning of 2021-07-14 (Radio Hagen 2021). The Nahmerbach flows N, in a valley east of the centre of Hagen, and joins the Lenne, a left tributary of the Ruhr, some 7km E of Hagen town centre. A road runs the length of the valley, where there was significant destruction.

---

[8] Hagen apparently has no weather station belonging to the German Weather Service network, but it does have its own network of measuring stations (Wetternetz Hagen, no date) .

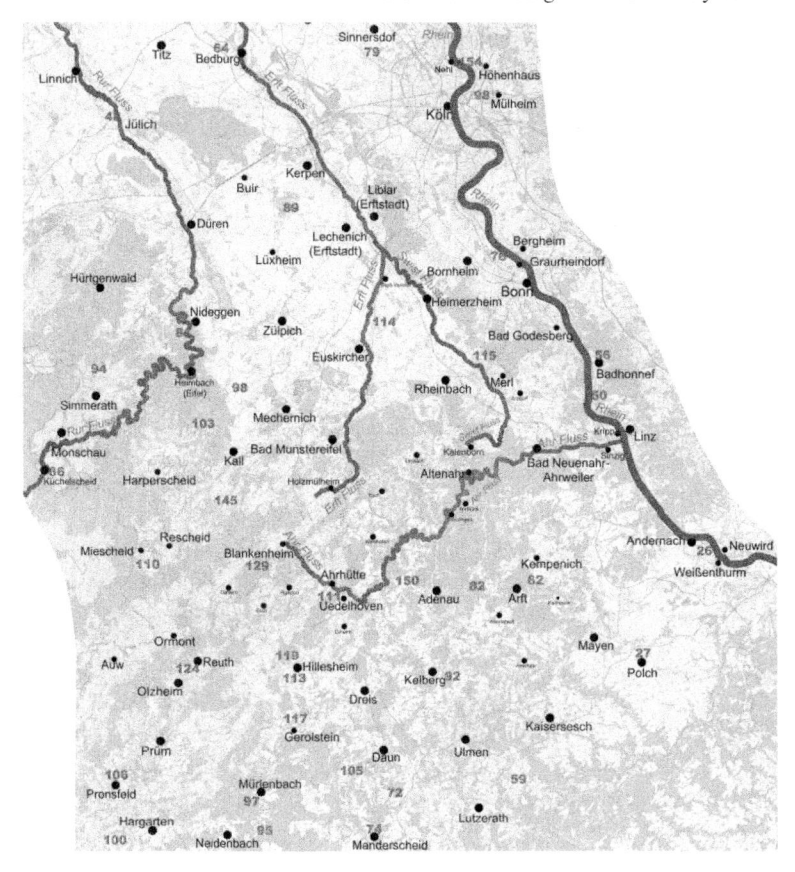

Fig. 14. The Ahr, Erft and Rur Watersheds, with Rainfall levels in the Eifel Massif (Credit: Lou Xinxin, OpenStreetMap)

3.2 Weight, Mass and Volume

In 24 hours, Hagen-Nahmerbach received *1/5 of a tonne of water per square metre*[9], and some places in the Eifel, as well Köln-Stammheim, about *1/6 of a tonne per square metre*. Many places in the same general area had 1/10 of a tonne per square metre. 150 mm = 15 cm ≈ 6 inches. On one square metre, such a depth of water weighs 150 kg, 1/6 of a tonne, more than one but less than two people.

---

[9] It is well to note what this means physically in terms of amounts and weight of water and possible force. One cubic metre of water weighs one tonne, 1,000 kg. 1 litre of water weighs 1 kg (at standard temperature and pressure and isotopic composition). It is 1mm deep when spread out over 1 m². So a depth measurement of mm/m² can equally be thought of as an amount measurement (in litres per square metre) or a weight measurement (kilograms per square metre).

It does not take much depth of flowing water, say 15 cm, to sweep you off your feet (especially if the ground is uneven and you can't see it), as walkers fording mountain streams know. In a mountain stream, you can see where the water is going and you can mostly see the bottom. In contrast, turbid water flowing rapidly and unpredictably in floods on uneven ground is dangerous stuff.

Suppose 100 mm/m^2 fell on most of the Ahr watershed, 900 km^2 of it. 900 km^2 is 900,000,000 m^2. That amounts to 90 billion litres of water, or 90 million tonnes, quite a lot of it draining into the river (estimates vary; Magnusson et al 2021 suggest 20-25% is "*likely to be an underestimation*").

4 Belgium

There were over 40 deaths in Belgium (Wikipedia-2021-Hochwasser, Belga 2021) and much physical damage.

The towns and cities affected in Belgium were mostly along the watershed of the Hohes Venn, in particular along the valley of the Vesdre/Weser. The Vesdre flows from east of Eupen, where there is a lake and a dam, the Wesertalsperre (Vesdre Valley Dam) which may be seen in Figure 1, and continues through quite a narrow valley in the Ardennes to Verviers and Pepinster, flows into the Ourthe near the eastern suburbs of Liège, which then flows into the Meuse in Liège itself (Wikipedia-Vesdre, no date).

The area around Eupen is shown in more detail in Figure 15. Near the crossing yellow roads in the centre of the city, the river Hill (Fr. Helle) flows north into the Weser (Wikipedia-Hill, no date). The mayor of Eupen, Claudia Niessen, said that the waters of the Hill can in fact be fed into the Wesertalsperre, but this evidently had not happened (Dendooven 2021). There was considerable debris brought down with the flow of the Hill.

The Corman milk-products factory in Baelen (Figure 16, just west of Eupen) is said to use about 10% of Belgian milk production, and makes Balade butter. The managing director, Vincent Mazy, said they had built a 4.5-metre dyke in anticipation of possible floods, and they have two hectares of land which normally suffices to allow floods to spread out, but the factory was hit by "two tsunamis" in the night, which swept away all material on the factory floor, including large blocks of fat weighing a tonne each in storage, and parts of the factory building itself (*op. cit.*).

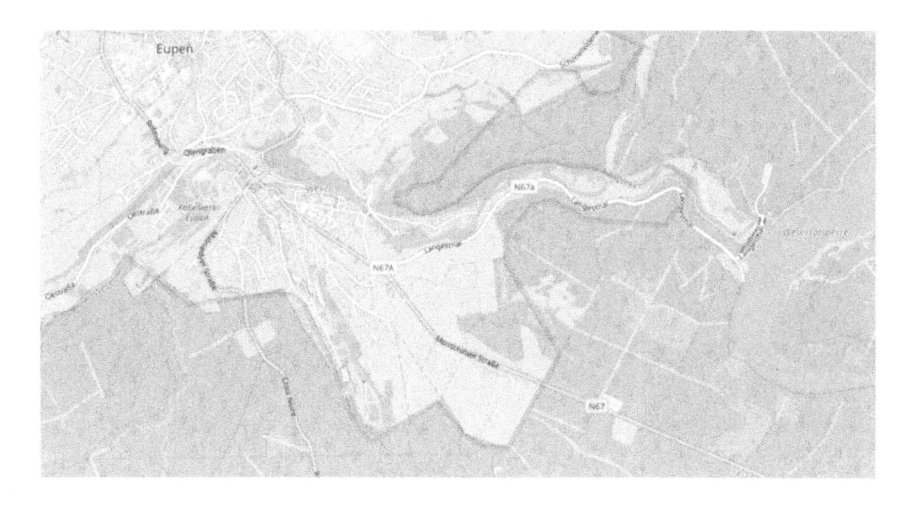

Fig. 15. Eupen and the Wesertalsperre (Vesdre Valley Dam)
(Credit: OpenStreetMap)

Figure 16 shows the railway line[10] coming from Aachen towards Verviers and further to Pepinster and on to Liège. It follows the valley of the Vesdre from the town of Dolhain/Limbourg. The line is somewhat higher than the river and the road, and you are often looking down onto the small towns. Figure 17 shows the Vesdre flowing through Limbourg in more placid times. A similar view may be seen from the train, which passes over a viaduct a little upstream.

The destruction was perhaps heaviest in the town of Pepinster, just downstream (west) of Verviers. Pepinster lies at the confluence of the Vesdre with the Hoëgne, another northwards-flowing river (in Pepinster) with its origins in the Hohes Venn. The mayor, Catherine Delcourt, said 59 buildings were damaged, including about 40 houses, and 350 people have had to be rehoused (C.H. 2021).

---

[10] The line used to be the slow route of the high-speed Thalys and ICE Cologne-Brussels trains, through Aachen and then Liège, until the ETMS line came into service on the high plateau between Aachen and Liège. It is twisty and pretty.

Fig. 16. The Vesdre, with Eupen, Baelen, Verviers and Pepinster
(Credit: OpenStreetMap)

Where did this come from? Again, the topography is narrow river valleys; but this time with river confluences in Eupen and Pepinster. Most of this water originates in the Hohes Venn. But the rainfall there was not high in comparison with other parts of the Eifel, or in comparison with Hagen. Aachen recorded 55 mm/m^2 on 2021-07-13 (from 0600 to 0600 on the morning of 2021-07-14) followed by 97 mm/m^2 on 2021-07-14 (again from 0600 for 24 hours). Monschau, just off the E edge of the Hohes Venn, recorded 17.2 mm/m^2 on 2021-07-13 and 65.5 mm/m^2 on 2021-07-14. The Hohes Venn itself contains a lot of water in any case, so it may well be that a large part of the inundation on 2021-07-14 turned into runoff down the draining rivers.

Fig. 17. The Vesdre in Limbourg. Credit Jean-Pol Grandmont (CC BY-SA 3.0)

There was concern expressed in the local news in Eupen that the Wesertalsperre was full, and speculation that some 7.3m cubic metres of water could have been let out, from Monday 2021-07-12 onwards, after the threat of heavy rainfall became known. According to some judgements, it would not have avoided the flooding but it might well have mitigated the consequences (BRF 2021), because water had to be released at the time of the flooding on 2021-07-14 to prevent an overrun[11] of the dam.

The head of the reservoir system for Wallonie, Fabian Docquier, is reported (op. cit.) as saying to the newspaper La Meuse that they were warning on Monday, but only informed Tuesday of the amount of rainfall expected, and at the point they didn't want to risk flooding the valleys themselves with a massive release (presumably accompanied by heavy rainfall and runoff). And then more rain fell than (someone? they?) had anticipated. Warning was available more than two days in advance and certainly with estimates of the amount available well more than the day before. There is surely scope here for "joining things up".

The Belgian German-language news station BRF also reported (BRF op. cit.) that a spokesperson for the Wallonie reservoir system had told them on Thursday 2021-07-15 that you couldn't really anticipate such events that only occur every 50 years. That seems odd. Of course you can plan for anything that happens within living memory, or even within the historical record. I suspect something else was meant, maybe as related by the news weatherman Sven Plöger, reported in the same article to have said that, yes, 150-200 mm/m^2 had been mentioned, but for example the prediction for Aachen had been 20-150 mm/m^2; and that it does not necessarily make sense always to react to extremal values of estimates. Plöger noted that reacting to given extremal values would lead to many false warnings, which would in turn lead to people and organisations becoming complacent, which in itself is dangerous.

5 The Weather System

The cause of all this was known some four days before, at least. Unusually heavy rain had been forecast by the European Centre for Medium Range Weather Forecasts at least 72 hours in advance, in Figure 18. This was taken on board by the German Weather Service, as well as other weather prognostication sites (e.g. Ruhnau 2021).

---

[11] An overrun is different from an overflow. A dammed lake has a designated maximal level, lower obviously than the top of the dam. An overflow occurs when this maximal level is reached. Overflows are provided for and accommodated in the design of the dam, with a spillway and other devices. I am using the term "overrun" to mean when the water level is actually higher than the top of the dam and water flows uncontrollably over the top.

The general weather situation was as follows. There was very moist, moderately warm air over much of this area, capped by a "blob" of colder upper air. Such upper-air systems are usually moved along by/with the jet stream, or dispersed by it over a few days. But the jet stream was somewhere else. This has happened at times in recent years. The upper air blob was not moving.

If the air below had been sufficiently warm, say, some 5+° warmer than it was on 2021-07-14, strong convection could have resulted. Such convection cells form thunderstorms and with thunderstorms comes, yes, rain, but also considerable wind and atmospheric disturbance in general, which tends to keep things moving. Thunderstorms are transient and relatively local phenomena, which do not generate persistent, heavy rain for hours or days on end, as happened in this event.

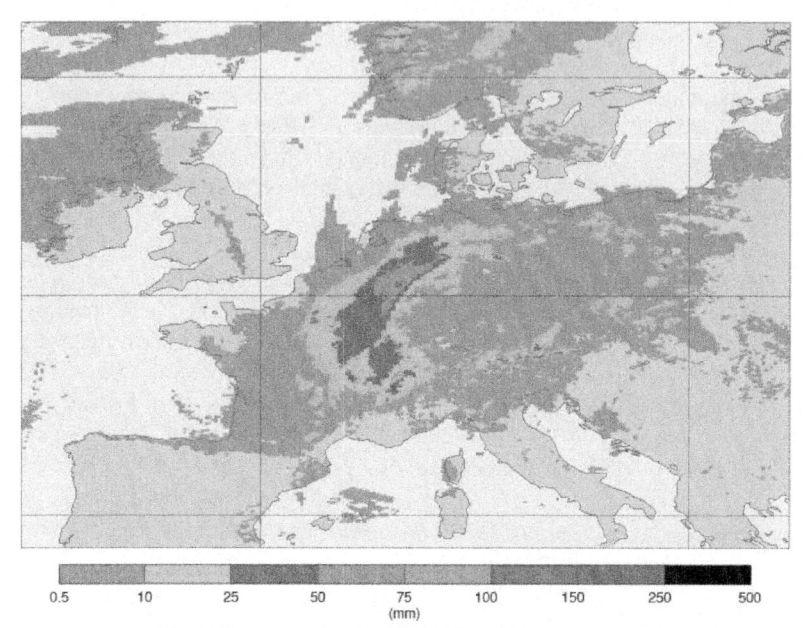

Fig. 18. The 72-hour Forecast of Rainfall for 2021-07-14 (ECMWF 2021)
Credit: European Centre for Medium-Range Weather Forecasts

This quasi-stationary system was quite extensive. It also covered us in Bielefeld, some 100km NE of Hagen, where the most fell, and some 200 km away NE as the crow flies from Köln-Stammheim, where the most fell outside of Hagen.

What is not really known beforehand is where such rainfall is likely to be worst, as can be seen by comparing the 72-hour forecast, Figure 18, with the 6-hour forecast, Figure 12 left frame, and the actual amounts, Figure 12 right frame and Figure 14. It could have happened over us in Bielefeld. But it didn't. The weather station 2 km down the road (and down the hill) from me, in Bielefeld-

Deppendorf, recorded a relatively miserly 10.4 mm/m², from the records displayed at (Wetterkontor, no date)). We had a rainy day, but nothing particularly remarkable.

How much moisture is in such an air mass, and how high is it? A post on the American Geophysical Union blog (O'Hanlon 2013) suggests that "most" of the European continent's extreme rainfalls from 1979 until that date were caused by atmospheric rivers, referring to (Lavers 2013). However, I have not heard the term "atmospheric river" in connection with the weather events of 2021-07-14.

Because of their power, and that they do occur in Europe, any prophylactic measures against flooding must surely consider atmospheric rivers, even if one was not involved in the events of 2021-07-14. The atmospheric river known as the "Pineapple Express" discharged on the Canadian province of British Columbia just before the time of writing, and appears to have shed some 600 mm/m² rain in a couple of days on some parts east of Vancouver (Watson et al 2021). For comparison, this is equivalent to two days of the highest recorded rainfall ever in Germany, in Zinnwald in 2002 (Freistaat Sachsen 2004).

Atmospheric rivers are rivers of moist air, typically flowing at an altitude of 10,000ft to 15,000 ft, and they can carry about 20 times the flow of the Mississippi river (Ramirez 2021). The average discharge of the Mississippi is just under 18,500 m³/s according to Wikipedia (Wikipedia-rivers-discharge, no date), so 20 times this flow is 370,000 m³/s, about 1m tonnes every 3 seconds. Unlike land-based waterways, though, the width of a typical atmospheric river is up to hundreds of miles.

The weather system causing the rain of 2021-07-14 extended west into Belgium as well as east into hilly country east of Cologne called the Sauerland, the source of the Ruhr as well as the site of the large Möhne dam and lake. Hagen, on the northwestern border of the Sauerland, suffered damage as described above. Wuppertal, west of Hagen, lines a long, narrow and steep valley and suffered some, but not that badly – they temporarily lost an elephant sculpture of some many tonnes, placed in the middle of the River Wupper[12].

The European Flood Awareness System (EFAS) put out the following 10-day flood warning on 2021-07-12, in Figure 19. The red squares in Figure 19 are points of interest ('reporting points') on the river network where the flood signal was predicted to exceed the 1-in-5-year flood return period threshold. The river discharge hydrograph (inset) is shown for the River Meuse near Visé in Belgium (drainage area of 20,825 km²). Visé is 10km downstream of Liège and 10km upstream of Maastricht on the Meuse ("Maas" in Dutch). Notice the reporting points on the Meuse and on the Rhine, and the lack of such points on smaller

---

[12] The sculpture is of an elephant named Tuffi, who in 1950 was being transported on the suspension railway hanging over the river, and who jumped out of the car and landed on its backside in the mud, unharmed (NW 2021.1). The sculpture is basalt, was washed off its pedestal, and found next to it some time later when the waters had receded (Wuppertaler Rundschau 2021).

rivers such as the Ahr, Erft and of course the Swist and Nahmer, which are called streams ("Bach").

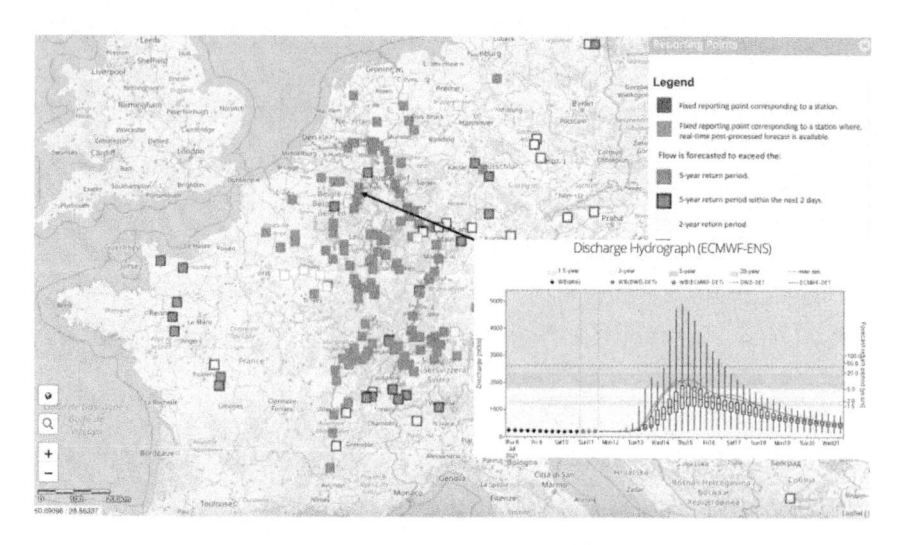

Fig. 19. EFAS 10-day Flood Warning from 2021-07-12 (ECMWF 2021)
Credit: European Centre for Medium-Range Weather Forecasts

6 Why?

So much for what happened. But with 180+ people having died, questions arise.

Note that the situation is unprecedented, not only in historical records but also in estimates. Preliminary assessments of the Erft flooding are apparently being carried out by an engineering firm at the behest of the Bezirksregierung Köln. Assessments have been previously performed for "1,000-year" events and apparently this one significantly overshot that estimate (NW 2021.2).

The "person of the day" in my local paper, the Neue Westfalische (NW) on 2021-07-20 was a hydrologist at the University of Reading named Hannah Cloke. The newspaper noted that clear warnings in the European Flood Awareness System (EFAS) four days before the events were unheeded by the population. Professor Cloke suggested to the Science Media Centre that EFAS worked fine but there had been a "failure of action" when the warnings didn't lead to people getting out of the way (Science Media Centre 2021). News reports showed people wading around in flood waters when they shouldn't have been. As noted above, it doesn't take much to sweep you away.

A similar situation arose in consequence of the Asian tsunami of 2004. An early warning system was set up in the Indian Ocean, by, amongst others, consultants from the University of Potsdam. The early warning system delivered tsunami warnings – to centrally selected government points. The question of how to get such warnings back out, quickly, to the people in the coastal communities likely to be hit was initially inadequately addressed[13].

Armin Schuster heads the Federal Office of Civil Protection and Disaster Assistance (Bundesamt für Bevölkerungsschutz und Katastrophenhilfe, BBK). He was quoted in the news on 2021-07-19 (NW 2021.3). The warning infrastructure, he said, had "fully functioned". He said it "was not our problem, rather [the problem was] how officials reacted, and also the public" (my translation). The Federal Academy for Civil Defence (BABZ) is in fact located in Bad Neuenahr, one of the worst-hit towns (BBK, no date).

Contentious discussions have commenced. According to a report in the NW for 2021-08-02, the Frankfurter Allgemeine Zeitung (FAZ) reported that emails from the state administration of RP to the Ahrweiler District administration were sent on 2021-07-14, but apparently not reacted to (NW 2021.4, see also Schaible 2021). In the afternoon of 2021-07-14 they reported a high-water estimate of 3.7 m forthcoming on the Ahr. Later in the evening, the prognosis was for high water of 7 m by 21.30. Apparently the District administration only announced a catastrophic situation at 23.00 that evening, and an evacuation order was issued at 23.15, by which time apparently some houses had already been destroyed. All that may (or may not) be so, but what are people doing relying on email and WWW pages for urgent communication over the "last mile" in a life-threatening situation? One needs at least a communication system with error-detection and a positive-acknowledgement protocol.

On 2021-08-06, the NW reported comments by President Marian Wendt of the Association of Technical Assistance Organisations (NW 2021.5). The Federal Agency for Technical Assistance (Bundesanstalt Technisches Hilfswerk, THW) is the primary engineering civil defence organisation[14]. They have lots of heavy equipment. In case of floods, the THW builds sandbag walls, repairs and support dykes, and in Bad-Neuenahr in the Ahr valley they quickly erected a iron-girder bridge to replace one that was washed away, as well as rebuilding the approaches, to enable road traffic to cross the river once more. They help with evacuations (it often helps to have seriously heavy kit that can drive through certain levels of flooding).

---

[13] My then-student I Made Wiryana, now at Gunadarma University in Jakarta, was contracted by the Indonesian government to design, program and supervise the implementation of this warning-distribution system, which he did using an agile-design technique of his own invention derived from WBA-type analyses. His report on how this was done is part of his PhD thesis (Wiryana 2009).

[14] 99% of THW personnel are volunteer; a fair proportion are young people between school and university on a year's voluntary service (the service that replaced the compulsory draft on 2011-03-01). All their kit is bright blue, contrasting with the red of the emergency services.

Herr Wendt said that you couldn't hinder the sheer power of the flood and also not save all who died, but (echoing Professor Cloke) it had been possible to bring many more people to safety had the responsible politicians acted differently (my translation). He noted the delayed warning, mentioned above, in the Ahrweiler district, from which the most victims came. He said that a "different, centralised system" was needed to convey these warnings (from the BBK, and from the German Weather Service).

As we have seen, these warnings are currently conveyed to politicians and politically appointed local administrators, who are "responsible" but not necessarily experienced in such civil defence matters. Herr Wendt is basically saying that joined-up professional management is needed all the way down (my paraphrasing).

Let us be clear on the concept. An early-warning system produces messages. These messages are to be read and heeded by the recipients. The recipients of a flood warning system, or indeed any natural-disaster warning system are surely: (a) those who would be directly affected and (b) those who can directly help them (protection services, evacuation, the logistics supporting these activities). The system does not function if groups (a) and (b) do not get the warnings. To say that the system functioned but the latter two groups did not get the message is a contradiction. The story above suggests that (b) was not necessarily ensured, and that (a) was only ensured via (b). One can surely think hard about changing the system design.

One issue is that there are social and political boundaries and hierarchies. The responsibility and the organisation of relief are legally actions at the level of the individual German states (Bundesland), whereas the Federal emergency management represented by BBK and Herr Schuster, and the THW and Herr Wendt, is Federal (Bund). At some point in warning systems/responsibilities and so on, the Fed stops and the particular State takes over, as a matter of constitutional law. This entails that the system, the end-to-end joined-up system, is hybrid. Professor Cloke is correct to observe that it wasn't working optimally. The informational-communication issue surely needs addressing. It is heartening to see that politicians as well as professionals have apparently taken this point on board, independent of party affiliation.

It is, however, somewhat disheartening to see the state attorney's office in Koblenz (the capital of RP) opening an investigation into the Ahrweiler district head of administration (Landrat) for negligence resulting in death (the German denotation is "fahrlässige Tötung"[15]) (NW 2021.6). The Landrat is a politician, chosen directly by district voters. The position comes with defined responsibilities, and if there is doubt as to whether they were fulfilled, the state attorney's office is required to investigate. It is not a matter of discretion (although one can argue

---

[15] "Fahrlässige Tötung" does not quite correspond with English legal concepts for unlawful/avoidable killing.

that it is a matter of discretion to perceive a possible dereliction of duty in the first place).

According to the NW on 2021-08-18, the RP Landrat has declared he is unable to continue with his duties due to illness (NW 2021.7). One can understand fully. From running a delightful tourist district to being investigated for responsibility for ~100 deaths is a startling life-change. There surely must be some doubt as to whether such a legal action is commensurate with the situation. This is a known issue with German law and accidents, cf (Ladkin 2007).

Constitutional change is possible, has occurred in other situations, but I doubt it should be necessary in this case. I would suggest that the minimal necessary is some kind of state-level professional management directly linked to and adequately mirroring what the BBK and THW are doing; somebody or -bodies who understand the physical situation in professional detail. State- and district-level politicians can surely remain nominally in charge; it is to be expected that they will follow their experts' advice in such matters; but the advice needs to be there, and to be accurate and timely; the knowledge delivery mechanisms have to work.

There is a logistical problem of where to station and where to send the resources, for instance, heavy vehicles; boots; helicopters. Emergency relief resources cannot effectively cover an area of thousands of square kilometres in advance, which is the area covered by the weather warnings. There are simply not enough resources to station them everywhere they might eventually be used. At some point, relief forces have to be told that it's been raining like heck for hours in some parts of the Eifel and in Hagen and Cologne, and just moderately in Ostwestfalen (where I am), and they then have to get over to Hagen, Cologne and the Eifel rapidly. Helicopters are of course highly mobile and rapid, but they cannot generally fly in severe rainstorms in mountainous areas at night. Organising the resources is a non-trivial real-time undertaking which can likely be optimised further but cannot be perfectly realised.

There are things we can do in advance. Within the existing infrastructure, we have noticed there have arisen questions about dam management. As has been seen, the question was raised in Belgium whether the Wesertalsperre could have been opened days before, and up to 7.3m cubic metres of water released, before the rains came. Had information been available to them more than a day in advance, it seems as if water could have been released, and the reservoir could then have mitigated downstream flow in the Vesdre/Weser and Helle/Hill during peak rain by storing some of it. On the east side of the Hohes Venn, the Rurtalsperre overflowed on Friday 2021-07-16, and water much further downstream caused dam breaks in Wassenberg. Again, could this have been managed better, with the existing infrastructure? The Rurtalsperre, administered by WVER (Wasserverband Eifel-Rur, a governmental water administration entity), abuts the districts of Aachen and Düren; Wassenberg is in the Heinsberg district. There are four distinct administrative dominions right there, who have to get on board any proposed solution.

Also in advance, we can surely plan to make better use of local knowledge. What rate of water can be coped with reasonably in Schuld on the Ahr? What would the flow look like on the sealed soil in town? How big is the upstream watershed? You can make a table of (rainfall rates x hours) → consequences (water level, for example). I wonder if there is one and whether some civil defence person on the spot has it? The same questions can surely be asked for Bad Münstereifel, Erftstadt, Euskirchen, Bad Neuenahr-Ahrweiler and so on.

It turns out that my city, Bielefeld, has done (some of) the work: the local newspaper NW published areas susceptible to flooding, along with maps of inundation, given a rainfall of 90mm in 1 hour. The city WWW site has an interactive map of runoff flows and susceptible areas. We have 500 times the number of residents as the very small town of Schuld (660 residents) and our resources for such things are correspondingly larger. One could well consider centrally financing equivalent studies for vulnerable places without the resources to do it for themselves.

One can also query the assumptions that underlie the Bielefeld city estimates. For example, we have had a pond dam break, uphill from my neighbour, covering her driveway and the road in mud, with the city having to come and clean it up. As far as I can tell, that pond was not identified as a particular source of concern on the flooding estimates. Second, the road adjacent to mine is the downflow of a watershed of many hectares of farmland (mixed cropland and horse meadows). When it rains heavily, say for a few minutes during a thunderstorm, water mixed with farmland earth flows down the road and onto the main road (which also slopes downhill), because the entrance to the main culvert gets clogged (as it was with Autumn leaves as I took Figure 20). The overground runoff of mud has been substantial enough at times to hinder traffic on the main road, but I am unaware that anybody in the city administration has ever talked to any of us residents about it. We are surely not the only people with a repeated clogged-culvert phenomenon. It is a circumstance calling for an engineering solution.

Fig. 20. The Culvert on Am Hobusch
(Credit: Author) The glove is next to it for comparison

There is lots of sealed soil around our houses, rather more than there was 20 years ago. The sealing stops a couple hundred metres downhill, and there is a wide stream below the road at that point. Storm runoff has therefore hardly ever been an issue, but then again very heavy rain has usually been associated with thunderstorms, and thereby of short duration.

There is also local knowledge not just in the form of what happens, but in the form of what can be done. I think there are plenty of people in my neighbourhood available for "volunteer civil defence", if such were organised. A simple example: who has portable sump pumps? I do. I don't think anybody else knows that. If there were to be a list of who has what kit and is willing to use it, then if someone's cellar should flood, a few of us with our pumps could well take care of it, and thereby relieve the fire service, who would otherwise come out. There are legal issues concerning responsibility which would need to be clarified via formal structures, such as a formal neighbourhood civil-defence organisation. Maybe we can take it up with the district council.

7 Could It Have Been Worse?

Certainly it could have been worse. The dam on the Steinbach, the Steinbachtalsperre, could have collapsed. Even more rain could have fallen (most of it was not the 200+ mm measured in Hagen, but over the stricken areas more like 50-100 mm). Higher rainfall probably would have turned the delayed overrun of the Rurtalsperre into an immediate threat, on the evening of the storm, to Jülich and Düren.

The amount of rain which fell in Hagen was just over 200 mm/m^2 in 24 hours, two thirds of the German (recorded) record at Zinnwald in 2002 (Freistaat Sachsen 2004). It is surely worth noting that, in Zhengzhou in Henan on 2021-07-17, that amount of rain, 201.9 mm/m^2, fell in just one hour (Wikipedia-2021-Henanfloods 2021). And, as noted above, rain in mid-November 2021 in parts of British Columbia accumulated to 600 mm/m^2 in two days (op. cit.).

8 Discussion

From the situation description and the discussion above, we can identify areas of possible improvement and prophylaxis.

1. Residents were not necessarily aware of what was coming, nor how to conduct themselves in this emergency. One could consider "flood defence" training for residents of topographically constrained landscapes such as the Ahr watershed.

2. Warnings of the impending event could have been transmitted to residents and local civil defence units earlier, and acted upon earlier. The current information path is complex and could evidently be optimised. There are complex sociotechnical structures in place, and warnings need to move through or tunnel through these structures rapidly. Technically, such messages would have error correction and be positively acknowledged upon reception – standard reliable-communications measures.

3. The logistical problem of where to station resources is very difficult, in particular given topographic constraints and the geographical uncertainty of actual precipitation amounts on the occasion.

4. Proactive dam management could not only prevent overruns but also provide capacity to retain some initial runoff.

5. Overrun dynamics in flood plains can be engineered. Along a constrained topography such as the Ahr watershed, this is likely not very possible, but for example north of the massif, in and around Euskirchen and beyond, there is flat or relatively flat land in which artificial flood plains could be constructed.

6. More precise topographical knowledge could be developed to mitigate extreme events. In valley towns, where exactly does runoff flow; in what order do inundations occur? Resources to form this knowledge for very small towns such as Schuld likely need to be implemented at the state level (at least).

7. Neighbourhood assistance systems could be developed. Who has what kit; when should it be mobilised? A local civil defence could be organised.

8. Last but by no means least: engineering simulations of such events and more extreme events can help planning and preparedness.

Six of these eight measures involve advances in engineering. Seven involve sociotechnical systems. *Sociotechnical systems engineering* has aspects which are not included in systems engineering as currently practiced, deriving from the human organisational structures (or lack of them). Such events as these floods and their consequences highlight the need for sociotechnical systems engineering.

Acknowledgments Many thanks to Dr.-Ing. Lou Xinxin, for the elegant Figure 14. Many thanks also to the Rhein-Erft-Kreis and its press officer Thomas Schweinsburg for permission to use the pictures in Figures 6 and 8. Thanks also to the reviewers for very helpful comments, which I have done my best to incorporate.

References

BBK (no date) Bundesamt für Bevölkerungsschutz und Katastrophenhilfe, Bundesakademie für Bewölkerungsschutz und Zivile Verteidigung, https://www.bbk.bund.de/DE/Themen/Akademie-BABZ/akademie-babz_node.html (in German)

Belga (2021) Belga, 38 morts, près de 100.000 personnes touchées : le lourd bilan des inondations égrainé au parlement wallon, RTBF 2021-09-01 https://www.rtbf.be/info/belgique/detail_38-morts-pres-de-100-000-personnes-touchees-le-lourd-bilan-des-inondations-egraine-au-parlement-wallon?id=10834063 (in French)

BPB (2021) Bunderzentrale für politische Bildung, Jahrhunderthochwasser 2021 in Deutschland https://www.bpb.de/politik/hintergrund-aktuell/337277/jahrhunderthochwasser-2021-in-deutschland

BRF (2021) BRF Nachrichten, Hochwasse: Hat man die Jahrhundertflut verpennt? 2021-07-19. https://brf.be/regional/1507853/ (in German)

C.H. (2021) C.H., Jusqu'à 59 bâtiments devront être détruits à Pepinster et 350 habitants relogés, RTBF, 2021-07-28. https://www.rtbf.be/info/regions/liege/detail_jusqu-a-59-batiments-devront-etre-detruits-a-pepinster-et-350-habitants-reloges?id=10813175 (in French)

Dendooven (2021) Dendooven L, Quel rôle le barrage d'Eupen a-t-il joué dans les inondations de la vallée de la Vesdre? Pourquoi ne pas avoir évacué? RTBF, 2021-07-23. https://www.rtbf.be/info/societe/detail_quel-role-le-barrage-d-eupen-a-t-il-joue-dans-les-inondations-de-la-vallee-de-la-vesdre-pourquoi-ne-pas-avoir-evacue?id=10810019 (in French)

ECMWF (2021) European Centre for Medium-Range Weather Forecasting, Heat, rain, floods and fires – the European summer of 2021, 2021-09-07 https://www.ecmwf.int/en/about/media-centre/news/2021/heat-rain-floods-and-fires-european-summer-2021

ECMWF (no date) ECMWF Reanalysis v5 (ERA5) https://www.ecmwf.int/en/forecasts/dataset/ecmwf-reanalysis-v5

Freistaat Sachsen (2004) Das Lebensministerium des Freistaats Sachsen, Ereignisanalyse: Hochwasser Augus 2002 in den Osterzgebirgsflüssen, Juli 2004 https://www.umwelt.sachsen.de/umwelt/infosysteme/lhwz/download/Ereignisanalyse_neu.pdf

GA-Pictures-of-Ahr (2021) General Anzeiger, https://ga.de/fotos/region/flut-entlang-der-rotweinstrasse-an-der-ahr-bilder_bid-62152563 (in German)

GA-Pictures-of Swisttal (2021) General Anzeiger, https://ga.de/fotos/region/ueberschwemmungen-erftstadt-rheinbach-swisttal-euskirchen-bilder_bid-61473809 (in German)

Kalischek (2021) Kalischek I, Was die Flutopfer wütend macht, Neue Westfalische Zeitung, 2021-08-13 (in German)

Knaack (2021) Knaack T, Hochwasser und Überschwemmungen in Düren und Jülich – Damm übergelaufen, Südwest Presse, 2021-07-19. https://www.swp.de/panorama/rurtalsperre-eifel-hochwasser-ueberschwemmungen-aktuell-heute-dueren-juelich-evakuierung-sperrungchen-koennen-nach-juelich-zurueck-58212213.html (in German)

Ladkin (2007) Ladkin PB, Negotiating Accidents: Analysis and Blame, Ninth Bieleschweig Workshop, Hamburg, May 2007 https://rvs-bi.de/Bieleschweig/ninth/LadkinB9Slides.pdf

Lavers (2013) Lavers DA and Villarini G, The nexus between atmospheric rivers and extreme precipitation across Europe, Geophysical Research Letters 40(12), 2013-06-28, doi: 10.1002/grl.50636 https://agupubs.onlinelibrary.wiley.com/doi/full/10.1002/grl.50636

Magnusson et al (2021) Magnusson L, Simmons A, Harrigan S and Pappenberger F, Extreme Rain in Germany and Belgium in July 2021, Newsletter 169, European Centre for Medium Range Weather Forecasts, https://www.ecmwf.int/en/newsletter/169/news/extreme-rain-germany-and-belgium-july-2021

NW (2021.1) Neue Westfalische Zeitung, Wuppertal sucht Elefanten "Tuffi", 2021-07-20 (in German)

NW (2021.2) Neue Westfalische Zeitung, Flut an der Erft soll simuliert werden, 2021-10-11 (in German)

NW (2021.3) Neue Westfalische Zeitung, Problemfall Katastrophenschutz: Streit und Schuldzuweisungen, 2021-07-19. https://www.nw.de/nachrichten/nachrichten/23053567_Wissenschaftlerin-wirft-Behoerden-monumentales-Versagen-vor.html (in German, paywall)

NW (2021.4) Neue Westfalische Zeitung, Landkreis missachtet Warnungen, 2021-08-02 (in German)

NW (2021.5) Neue Westfalische Zeitung, THW-Chef wirft Landräten Versagen bei Flutwarnung vor, 2021-08-02 (in German)

NW (2021.6) Neue Westfalische Zeitung Ermittlungen gegen Landrat im Ahrtal, 2021-08-07 (in German)

NW (2021.7) Neue Westfalische Zeitung, Hochwasser: Landrat lässt Amt ruhen, 2021-08-18 (in German)

O'Hanlon (2013) O'Hanlon L, Most of Europe's extreme rains causes by 'rivers' in the atmosphere, American Geiohysical Union blog, 2013-07-19. https://blogs.agu.org/geospace/2013/07/19/most-of-europes-extreme-rains-caused-by-rivers-in-the-atmosphere/

Radio Hagen (2021) Radio Hagen, Das Unwetter und die Folgen, 2021-07-14 https://www.radiohagen.de/artikel/das-unwetter-und-die-folgen-1007679.html (in German)

Ramirez (2021) Ramirez R, Weather Whiplash, CNN 2021-10-23 https://edition.cnn.com/2021/10/23/weather/climate-change-atmospheric-river-california-drought/index.html

Ruhnau (2021) Ruhnau F, Meteorologische Chronologie der Flutkatastrophe im Westen Deutschlands im Juli 2021, 2021-07-19, Kachelmannwetter.com ttps://wetterkanal.kachelmannwetter.com/meteorologische-chronologie-der-flutkatastrophe-im-westen-deutschlands-im-juli-2021/

Schaible (2021) Schaible I, Reichert B and Bauaer M, Die zerstörerische Nacht im Ahrtal, Neue Westfalische Zeitung 2021-08-07 (in German)

Science Media Centre (2021) expert reaction to flooding in northern Europe, 2021-07-15 https://www.sciencemediacentre.org/expert-reaction-to-flooding-in-northern-europe/

Stegmann (2021) Stephan Stegmann, Bagger-Fahrer rettete Steinbachtalsperre, General Anzeiger, 2021-07-10 https://ga.de/region/koeln-und-rheinland/wenn-die-wand-gebrochen-waere-haette-ich-keine-chance-gehabt_aid-61678349 (in German)

Swisttal (no date) Gemeinde Swisttal https://www.swisttal.de (in German)

Swisttalrat (no date) https://www.swisttal.de/cms125/gemeinde_rat_verwaltung/politik/rat/ (in German)

Tagesschau (2021) Tagesschau news, Liveblog, 2021-07-17 https://www.tagesschau.de/newsticker/liveblog-hochwasser-103.html (in German)

Watson et al (2021) Watson C, Sheehy F and Guest P, How bad is the pacific north-west flooding and what caused it? The Guardian 2021-11-18 https://www.theguardian.com/environment/2021/nov/17/pacific-north-west-flooding-british-columbia-washington-state-canada

Westbrock (2020) Westbrock S, Hinter der Kulissen der Steinbachtalsperre, General Anzeiger, 2020-08-18. https://ga.de/region/voreifel-und-vorgebirge/hinter-den-kulissen-der-steinbachtalsperre_aid-52838901 (in German)

Wetterkontor (no date) Historical Weather Data https://www.wetterkontor.de/wetter-rueckblick/ (in German)

Wetternetz Hagen (no date) https://wetternetz-hagen.de (in German)

Wikipedia-2021-European-floods (2021) https://en.wikipedia.org/wiki/2021_European_floods

Wikipedia-2021-Hochwasser (2021) Hochwasser in West- und Mitteleuropa 2021 https://de.wikipedia.org/wiki/Hochwasser_in_West-_und_Mitteleuropa_2021

Wikipedia-2021-Henan-floods (2021) https://en.wikipedia.org/wiki/2021_Henan_floods

Wikipedia-Ahr (no date) https://en.wikipedia.org/wiki/Ahr (in German)

Wikipedia-Hill (no date) https://en.wikipedia.org/wiki/Hill_(stream)

Wikipedia-rivers-discharge (no date) https://en.wikipedia.org/wiki/List_of_rivers_by_discharge

Wikipedia-Rur (no date) https://en.wikipedia.org/wiki/Rur

Wikipedia-Steinbachtalsperre (2020) https://de.wikipedia.org/wiki/Steinbachtalsperre_(Nordrhein-Westfalen)

Wikipedia-Swist (no date) https://dewiki.de/Lexikon/Swist (in German)

Wikipedia-Vesdre (no date) https://en.wikipedia.org/wiki/Vesdre

Wiryana (2009) Wiryana IM, A Sustainable System Development Method with Applications, Ph.D. Thesis, Bielefeld University. https://rvs-bi.de/publications/Theses/Thesis_2009_IMadeWiryana.pdf

Wuppertaler Rundschau (2021) Tuffi ist nach dem Hochwasser zurückgekehrt, 2021-09-24 https://www.wuppertaler-rundschau.de/lokales/stoerstein-in-der-wupper-tuffi-ist-nach-dem-hochwasser-zurueckkehrt_aid-62900843 (in German)

Introducing a Restorative Just Culture and the Learning Review at the Docklands Light Railway

Adam Johns

    KeolisAmey Docklands

Abstract *In 2020, KeolisAmey Docklands (KAD) – franchise operator of the Docklands Light Railway (DLR) in London – commenced work on a pioneering safety programme called 'Next Platform'. In what is thought to be a first for the UK rail industry, KAD are seeking to radically change the way they learn from adverse events and everyday work. Core to this transformation is the introduction of a 'restorative just culture', and a new safety investigation methodology called the 'Learning Review'. This new approach is as much about changing philosophy as it is about changing process. In a Learning Review, in which a restorative just culture process is embedded, the decisions and actions of staff involved in adverse events are reviewed in a neutral and curious manner to understand why they made sense to the person at the time. All inquiries are approached from a presumption of good intention, staff are treated with support and care, and as long as no intentional wrongdoing is identified, everyone's focus – leaders, managers and frontline staff – is on learning and improving. This talk will explore how KAD developed the Next Platform programme, why they decided to take the approach, and how it has benefitted the DLR operation so far.*

© Adam Johns 2022.
Published by the Safety-Critical Systems Club. All Rights Reserved.

Human Reliability in Complex Systems

Rachel Selfe

Atkins Ltd, member of the SNC-Lavalin Group

Abstract *Command and Control (C2) systems form an integral part of military defence in supporting operational military decision-making. C2 systems are complex socio-technical systems made up of physical hardware, software, data, Operators, Maintainers, processes and the workplace. One of the complexities is multiple Operators performing different roles simultaneously, under high workload, making rapid decisions to work towards a common goal. This paper outlines an ongoing project to replace an existing C2 system. A Human Reliability Assessment (HRA) is being conducted with an approach which incorporates a more holistic assessment of the human contribution within 'the system'. Identifying not only opportunities to identify and mitigate potential human error by influencing design but also looking at the wider interactions within the system and identification of emergent properties which may result from more complex interactions within the system.*

1 Introduction

1.1 Overview

A project is being undertaken to develop a Command-and-Control System (C2). Human Factors (HF) are an integrated part of the project team informing the Development of the system with HF analysis and recommendations to influence the design of the system. The HF team are undertaking Human Reliability Assessment (HRA) on Operator Tasks aimed at improving reliability of the human elements of the system through influencing design of the system. This paper presents a systems-based approach towards HRA which goes further than more traditional methods. What existing approaches cannot do is assess human reliability more holistically within the context of the wider socio-technical system, especially systems such as C2 which are dynamic and Operator tasks cannot be exactly prescribed and written down in a detailed procedure. Traditional approaches do not adequately assess the impact of dependency within the system and the complexity of interactions (including emergent properties) which take place

© Atkins member of the SNC Lavalin Group.
Published by the Safety-Critical Systems Club. All Rights Reserved.

(Leveson and Thomas, 2018). Therefore, this paper presents the Complex-Systems HRA and the integration of the approach early enough in the design phase to be able to influence design of the system.

1.2 What is Command and Control?

Command and Control (C2) systems are the facilities, equipment, communications, operating procedures, and personnel essential for military Commanders to perform the function of command and control during military operations. Whilst there is no formal UK or North Atlantic Treaty Organisation (NATO) definition for C2. The following quote is useful when discussing C2 (MOD JCN 2/17, 2017). Within a military context, command has been described as:

> 'the authority and responsibility for effectively using available resources and for planning the employment of, organizing, directing, coordinating, and controlling military forces for the accomplishment of assigned missions. Success in command is impossible without control' (Field Manual 6-0, 2003).

Within command and control:

> 'control is the regulation of forces and battlefield operating systems to accomplish the mission in accordance with the commander's intent. It includes collecting, processing, displaying, storing, and disseminating relevant information for creating the common operational picture, and using information, primarily by the staff, during the operations process' (Field Manual 6-0, 2003).

In the world of C2, mission success is critical.

A key requirement of C2 systems is the need to provide Operators with a sufficient level of Situational Awareness (SA) to make mission critical decisions. SA is a term used to describe an Operator's dynamic understanding of 'what is going on' (Endsley, 1995a in Salmond, et al 2007). Maintaining SA for Operators to make complex decisions is a fundamental part of C2 systems.

1.3 What makes C2 systems so complex?

For a C2 system to work effectively and consistently, many elements of the system have to interact simultaneously, including multiple Operators performing different roles to achieve a common goal. C2 systems are complex socio-technical systems that combine core elements of people, tasks, technology and organisation as represented in Fig. . The core elements of the C2 include:

- **People** includes the Operators conducting their individual roles within a larger team, all with a common goal. Maintainers working simultaneously behind the scenes to monitor the system to ensure all aspects remain operational and conducting Maintenance where required. Other people which form part of the wider system include military personnel out in the field and outside Agencies. They interact with the Operators, providing information and support maintenance of SA.

- **Technology** makes up a significant part of the C2 system and includes the software that the Operators are interacting with, the communications technology, secure data communication, the system management software that the Maintainers use to monitor the status of the system and the physical hardware.
- **Tasks** includes the Standard Operating Procedures that the Operators and Maintainers are working to for their particular roles. Decision-making in task prioritisation and team contribution to complex tasks requiring coordinated effort of multiple Operators.
- **Organisation** includes the structure and the organisational design that can impact C2 effective performance and the organisational culture that can affect how an organisation functions. The culture of an organisation may also impact on the extent to which there is a shared sense of responsibility/purpose, responsiveness to C2 demands and positive elements such as camaraderie amongst Operators and Maintainers.

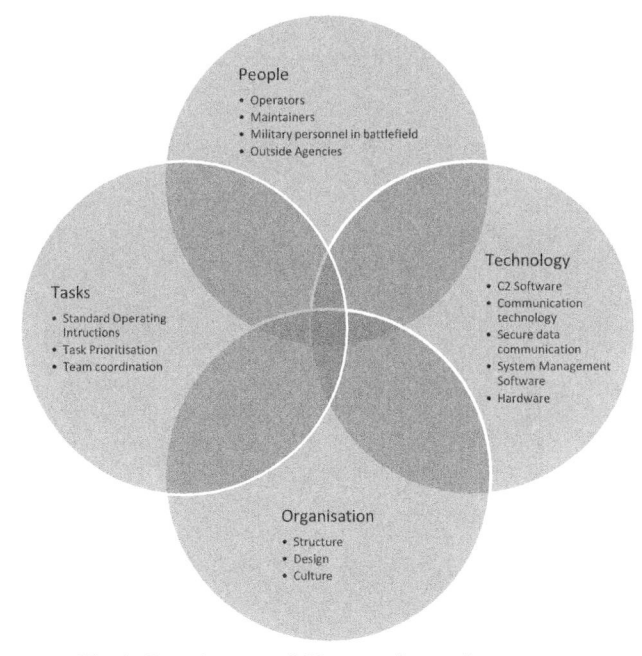

Fig. 1. Core elements of C2 system interacting

Figure 1 demonstrates the core elements and their interactions which make up a C2 system. Operators (people) dynamically interact with all elements of the C2 system (as displayed in figure 1) and are required to perform under high workloads to make rapid decisions. Therefore, C2 systems need to be designed to optimise Operator performance and minimise the likelihood for error. A key part of optimising that performance is ensuring that workload is kept to a minimum and SA is maintained to ensure appropriate decisions are made.

The complex interactions between individual components within a system such as C2 can create emergent properties (Leveson, 2016). Figure 1 demonstrates how the core elements of a socio-technical system can overlap and interact. In addition, figure 2 represents how individual components within a complex system interact. Some components interact directly with each other, and others interact indirectly.

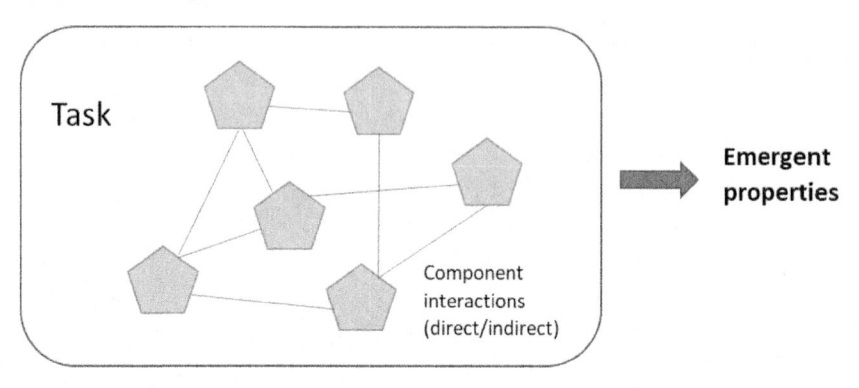

Fig. 2. The emergent properties which result from component interactions (Leveson, 2018)

Leverson (Leverson, 2018) explains that accidents can occur due to unsafe interactions among components that have not failed and that are still meeting requirements. One example provided by Leveson is quoted below:

> 'Some Navy aircraft were ferrying missiles from one point to another. One pilot executed a planned test by aiming at the aircraft in front (as he had been told to do) and firing a dummy missile. Apparently, nobody knew that the "smart" software was designed to substitute a different missile if that one was commanded to be fired was not in a good position. In this case, there was an antenna between the dummy missile and the target, so the software decided to fire a live missile located in a different (better) position instead.'

With this example provided it appears that no individual component part of that system failed. Instead, an unsafe emergent property arose as a result of the interaction between components.

Emergent Properties have also been described as '*characteristics of the whole are developed (emerge) from the interactions of their components in a non-apparent way*' (Bar Yam, 1997 in Walker et al, 2010). SA has been described as an emergent property within the context of C2 systems. Component failure and unsafe interactions can be controlled in a number of ways such as:

- Design (within the context of the C2 system – this can include design of the Human Machine Interface);
- Design of processes (Operational and Maintenance processes); and
- Social controls such as Regulation or Legal.

Leveson also highlights that organisational culture can also have an indirect control over emergent properties by affecting the conditions in which the tasks are performed, for example, the psychological safety among the team enabling perception that others will notify you if changes in SA occur or others will conduct the required supervisory checks.

Therefore, the types of components of the system that must be assessed to provide the appropriate level of safety assurance in C2 include: software, hardware, data/information, workplace, Operator tasks, Maintainer tasks, the organisation and its culture and procedures. One of the roles of the HF assessment is to aggregate the separate components of a complex system to conduct a holistic assessment of reliability of human performance. Therefore, HF being integrated into the design of a complex system is integral to the success of appropriately assessing reliability of human performance in a complex system such as C2.

A large element of the development of a C2 system focuses on the software functionality. One of the aims of HF is to ensure the functionality of the software aligns to the tasks the Operators are required to perform as part of their roles. This is because the software functionality and how it is designed and developed will affect how it is operated and consequently the performance of the Operators completing their task goals whilst using it. This requires a detailed assessment of the individual software components of the system. However, in order to capture those emergent properties resulting from the interactions between individual components of the system, HF must go further than traditional HRA approaches and consider the whole system and the wider interactions to conduct a more holistic assessment. These include interactions within the C2 software, those operating it, as well as face-to-face communications, radio communications with military personnel, outside agencies, other actors, and verbal and written shift handover communication.

This Complex-System HRA approach allows the emergent properties to be identified which may have been missed through more traditional methods that haven't taken a holistic approach to identify the system-wide interactions. In addition, conducting the HRA as the system is being developed allows the benefit of influencing the design as the work progresses.

1.4 Role of HRA in complex C2 development

The C2 system being designed and developed has been classed a safety-related system. As such, the approach adopted for this Complex-System HRA is a mixed method assessment of the human reliability with the C2 system. As documented within MOD Technical Guide 7-3 (MOD Technical Guide 7-3), the purpose of the HRA process is to:

- Ensure the design minimises the likelihood of human error and optimises human performance;
- Identify opportunities to mitigate human error through design (preferred) or procedural controls (where necessary);
- Document and provide the evidence required in the Safety Case that Operator and Maintainer tasks have been assessed, credible errors identified, and appropriate mitigation put in place and where claims are made on the human these are suitably substantiated.

1.5 Traditional HRA approach

Traditional HRA approaches break tasks down into their component elements, assessing them at their component level and in some cases creating fault tree models that illustrate the errors. This approach can be time consuming. However, if the HRA is focused on safety-related or safety-critical aspects of the system this can ensure a proportionate approach to HF is adopted. This more traditional approach is becoming less and less effective the more complex systems become.

The Complex-System HRA approach presented within this paper is going over and above more traditional approaches to HRA. Where more traditional approaches systematically analyse tasks sequentially through detailed task analysis, this is further expanded to include a system wide assessment to understand not just the task goals and sub-tasks but the wider elements of the C2 system and the dependencies which exist in achieving a task goal. Exploring interactions between the individual components as well to establish emergent properties.

As noted previously more traditional approaches cannot assess human reliability holistically within the context of the wider socio-technical system. Especially systems which are dynamic and Operator tasks cannot be exactly prescribed and written down in a detailed procedure. Traditional approaches do not adequately assess the impact of dependency within the system and the behaviours arising from complexity of interactions (emergent properties) which take place (Leveson and Thomas, 2018). Traditional HRA approaches focus on the human failure at each component level but not necessarily at the interactions between humans and technology.

C2 systems such as the one currently being developed are dynamic and Operator tasks and interactions with the system are scenario and context driven. As such a different approach to HRA is required to capture the dynamic interactions and decision-making taking place within the C2 system and subsequently the increased complexity of human error which needs to be identified and mitigated against.

1.6 Systems-Based Approach

As noted above C2 systems are complex socio-technical systems. As a result, the types of human errors which can occur are complex. Systems-Theoretic Process Analysis (STPA) is one type of hazard analysis technique designed to capture not only accidents which result from component failures, but also accidents which result from design flaws and unsafe interactions (Leveson, 2012). Whilst STPA is well-suited for analysing complex systems, it does not provide guidance specific to humans. An extension to the STPA method, "Engineering for Humans", was developed to provide guidance early in the design process and address human interactions in the system (Thomas and France, 2017 & France, 2017). This has gone on to be applied successfully in the aviation domain (Thomas and France, 2017). The approach presented within this paper aligns more with STPA and 'Engineering for Humans', by analysing the complex interactions between individual components of the system including Operators and identifying control actions to mitigate against emergent properties.

The following sections of this paper provides an overview of the ongoing Complex-Systems HRA approach currently being adopted to assess the human reliability within a project to develop a C2 system together with the benefits and limitations of the approach.

2 Method

2.1 Overview

The new C2 system requires a team of Operators to collect and analyse complex data before performing the appropriate response. Each Operator has a unique role to perform, conducting individual and, in some instances, overlapping tasks. An error made by one Operator could impact other members of the team and have more complex effects across the wider 'system'. The human contributions within safety assurance assesses all aspects of the socio-technical system.

The following details the mixed methods approach adopted to conduct the Complex-Systems HRA throughout the development of the C2 system. The approach is split into three separate stages as presented within figure 3. This includes:

- Assessment of software;
- Holistic Assessment of reliability; and
- User Trials.

The approach adopted and presented within this following section has been designed to ensure human interactions with the 'system' have been assessed appropriately in order to support the human reliability argument presented within the Safety Case. In addition, the approach has been designed to be undertaken throughout the development of the C2 system to highlight opportunities to improve human reliability within the system at the earliest opportunity and influence the design.

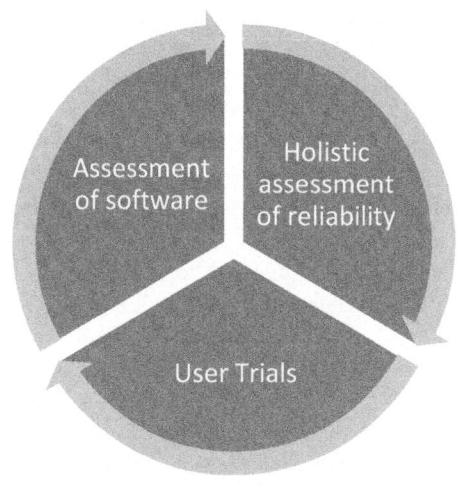

Fig. 3. Complex Systems HRA approach

2.2 Assessment of software

Whilst this paper has highlighted the importance of HRA in assessing the system as a whole, software and interface design forms a large part of the C2 system which the Operators interact with. As noted previously, maintaining SA to ensure decisions made by Operators are optimised, is critical in the development of C2 systems. The C2 interface is a critical component in maintaining SA for Operators. Having an interface which is designed well, is intuitive to use and is aligned with users' task goals can help to optimise Operators SA and consequently decision-making during missions. Therefore, optimising the usability through Heuristic reviews will help to improve how the Operator interacts with the system to optimise SA.

The purpose of the Heuristic reviews is to help ensure the interface is intuitive, usable, aligns with design principles, system requirements and user requirements and where possible highlights human error and makes suggestions on providing opportunity to recover from human error. A number of Heuristic reviews have taken place throughout the development phase of the system's interface. The reviews conducted have followed Nielson's Heuristic Principles (Nielsen, 1994). Each piece of new software functionality has been subject to a Heuristic review and where possible all elements of the functionality are assessed by HF. Recommendations have been made and fed back into the Development Team to influence the design of the interface.

2.3 Holistic assessment of human reliability

Holistic assessment of human reliability involves a range of tasks, starting with identifying new functionality to assess, through to end user validation. An overview of the holistic assessment of human reliability approach is presented in figure 4 and each step in the process will be described in more detail in the following sections.

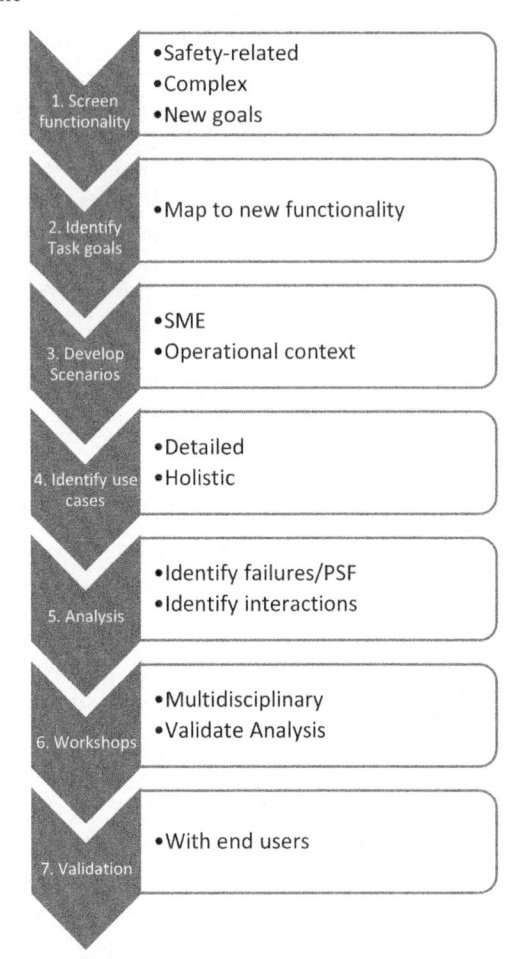

```
┌─────────────────┬──────────────────────────────┐
│ 1. Screen       │ • Safety-related             │
│ functionality   │ • Complex                    │
│                 │ • New goals                  │
├─────────────────┼──────────────────────────────┤
│ 2. Identify     │ • Map to new functionality   │
│ Task goals      │                              │
├─────────────────┼──────────────────────────────┤
│ 3. Develop      │ • SME                        │
│ Scenarios       │ • Operational context        │
├─────────────────┼──────────────────────────────┤
│ 4. Identify use │ • Detailed                   │
│ cases           │ • Holistic                   │
├─────────────────┼──────────────────────────────┤
│ 5. Analysis     │ • Identify failures/PSF      │
│                 │ • Identify interactions      │
├─────────────────┼──────────────────────────────┤
│ 6. Workshops    │ • Multidisciplinary          │
│                 │ • Validate Analysis          │
├─────────────────┼──────────────────────────────┤
│ 7. Validation   │ • With end users             │
└─────────────────┴──────────────────────────────┘
```

Fig 4. Systems-based approach to HRA

Task 1: Screen Functionality

Whilst holistic assessment was the aim of the HRA work conducted, the development of the C2 system has involved the addition of new functionality into an existing system. Functionality includes the command-and-control screens of the interface. These were identified through engagement with the Development team and the Human Machine Interface (HMI) Specification document. Therefore, the HRA was bound to focusing on the reliability of the tasks performed using the new functionality within the existing system. For a proportionate approach to be adopted, complex tasks undertaken by Operators and on the safety-related aspects of the system were identified via a task screening exercise. The purpose of the screening process was to filter out simpler and less complex functionality introduced by the new command-and-control screens. The following factors were considered during the screening process:

- Whether the functionality is safety-related (not all aspects of the system are classed as safety-related);
- Whether the functionality supports new tasks within the system;
- Complexity of Operator interaction with the functionality, such as amount of data entry, drop-down options, or read-only elements;
- Subjective judgement (HF assessor validated with Subject Matter Expert (SME) input) on opportunity for human error.

The output of the screening exercise was a reduced set of new functionalities to be taken forward to support further analysis.

Task 2: Identify Task goals
Task goals and sub-tasks undertaken by the human within the existing C2 system had previously been identified through a Hierarchical Task Analysis. New functionality identified during the screening exercise were then mapped onto one or more of the existing task goals. For example, a task goal would be 'Assign Mission'. Whilst most of the new functionality maps across to existing tasks, some are new aspects of the system which do not map across. Therefore, consulting with SMEs on how the new functionality would be used based on operational experience is valuable for understanding or identified any new task goals which have emerged via the introduction of the new functionality.

Task 3: Development of operational context scenarios
Once the task goals have been identified and the nature of the tasks understood, a set of scenarios were developed to aid the next step of holistic assessment. Detailed scenarios are developed in collaboration with SMEs which ensure the operational context is included when assessing the task-based goals. How Operators will interact with the C2 system will change depending on the operational context. Therefore, having a representational range of scenarios ensures that a robust assessment is conducted.

Task 4: Identify system wide use cases
For each goal identified the use cases to complete that goal are identified. Traditionally, use cases are developed within the Test Team detailing exactly how a 'user' will complete a task. Furthermore, this extended use case approach follows end-to-end requirements throughout the wider socio-technical 'system' beyond the functionality of the technical system alone. This extended approach is designed to detail the wider and on-going task goals such as maintenance of SA, as well as take account of the wider system. This identifies all aspects of the system needed to complete a goal and takes account of the varying options an Operator can take to complete a task goal. This process identifies all the direct interactions with the functionality, wider C2 system and the Operator including all the indirect interactions as well. The use cases detail both Operator and system inputs and outputs together with responses and actions required to

complete task goals. The use cases are developed in collaboration with SMEs and software Developers. The process of developing the use cases is the start of identifying emergent properties which can be built upon and validated at the multidisciplinary workshops.

Unlike traditional task analysis techniques, which identify all the task steps required to achieve the 'Operator goal, this (holistic) system-based approach identifies *all* aspects of the system which will support the Operator's successful completion of the goal. The aim is a more effective method of identifying not only the undesirable emergent properties but also the latent errors in the system and how they impact on the performance of the Operator in achieving the task goal.

Task 5: Analysis
Three stages of performance analysis are conducted and described below:
1. Identification of opportunities for error;
2. Component dependencies on other parts of the system;
3. Undesirable emergent properties.

Application of a Systematic Human Error Reduction and Prediction Approach (SHERPA) (Embrey, 1986) is used to identify opportunities for error within each use case. In addition to SHERPA, the analysis assesses each individual component to identify where use cases are dependent on other parts of the system (people and technology) and to identify latent errors and the consequences of latent errors if they remain within the system. Undesirable emergent properties such as loss of SA or other unsafe conditions are identified through the interaction of individual components of the system. Consideration is given to any direct (e.g. workload) or indirect (e.g. organisational culture) Performance Shaping Factors (PSF) which may impact on human behaviour and consequently onto performance of the Operators. The outputs from the analysis are a set of potential errors, key interactions or dependencies and potential undesirable emergent properties.

Task 6: Development and Validation Workshops
Multidisciplinary workshops are conducted to validate the findings of the analysis. SMEs, End-Users, software Developers, Safety and Human Factors attend to ensure a range of perspectives throughout the analysis. The workshops run through the new functionality, mapped to Operator task goals and use case analysis. The different operational context scenarios are discussed throughout the workshops to ensure all human errors, system interactions (dependencies) and emergent properties are identified and validated. The workshops give the opportunity for the Multidisciplinary team to develop mitigations for identified undesirable emergent properties.

Task 7: Validation of assumptions
The User Trials form a key part of the assessment. They form part of the validation of the assumptions made within the Complex-Systems HRA. The User Trials method and output use is discussed further in the section below.

2.4 User Trials

User feedback and validation such as User Trials form a key part of the user-centred design for the development of a complex system such as C2. In addition, User Trials form part of the overall HRA approach within this C2 project. User Trials provide an opportunity to validate any assumptions made during the previous HRA activity, to identify any human error not previously identified and to assess the interactions between the Operators and the System in achieving their common goals. User Trials are the first opportunity the End-Users have to use the new functionality of the C2 system within a representative control room environment. Simulated data is used within the system and exercises undertaken to assess task goals identified within the previous holistic assessment phase of HRA. Data gathered from the User Trials is from a mixed method approach including:

- Observational data exploring how users interact with the system, identifying where there is apparent confusion, hesitation and error;
- Error forms to explore the reasons behind any errors conducted;
- Semi-structured interviews to further explore any issues experienced during the User Trials and to further explore any errors.
- Wash-up sessions as a User Group to discuss issues identified as a group and suggested amendments to the system which may improve usability and Operator performance whilst using the system.
- Questionnaires which allow users to rate all aspects of the system.

Any issues identified with the system or new errors identified are fed back in to the HRA analysis to assess from a holistic perspective the wider impact on the system and task goals the Operators are performing. Recommendations are made to improve reliability. These are managed through the Human Factors Working Group[1].

---

[1] The Human Factors Working Group is made up of the key project representatives including the client, Subject Matter Experts, software Developers, Human Factors, Safety and End-Users. The group manage any risks and issues identified on the project through to closure. In addition, HF recommendations are presented to the group and decisions made to implement recommendations or not.

3 Discussion

3.1 Role of HRA as part of design

One of the key aspects of the approach presented within this paper is the integration of HF into the Development Team of the C2 interface. HRA is typically conducted at the end of a project and conducted on a completed design. However, this Complex-Systems HRA is implemented (combined with other HF techniques) to identify potential human errors and undesirable interactions early in order to influence design during the development phase of the C2 system. Wherever possible, designing out the opportunity for human error from a system is preferable over the reliance on training or procedural controls. These recommended changes are made if they are simple and require little in the way of development work for the software Developers. However, more complex recommended changes are presented to the Human Factors Working Group together with a cost/benefit argument to support decision making.

The benefits of this approach are that HF team are involved early on in the development of the Interface of the C2 system resulting in early design changes and fewer instances of reliance on training and procedural controls, which require Operators adjusting their behaviour and adopting compensatory trained behaviours rather than normal working behaviours in complex environments where C2 systems operate. Having HRA conducted once a system has already been designed and developed can result in an over reliance on training and procedural controls with less opportunity to identify and design-out potential human errors.

3.2 Benefits of Heuristics

Whilst Heuristic reviews are not formal HRA methods, the aim of the Heuristics is to improve the usability of the interface and consequently optimise the performance of Operators which use it. Conducting the Heuristic reviews throughout the development of the C2 system has identified numerous design modifications to the interface, including:

- Ensuring consistency throughout the interface such as labelling, and options presented within drop-down menus;
- Removal of unnecessary information to create a more simplistic and minimalist design;
- Positioning of buttons and tick boxes;
- Highlighting areas more prone to human error and suggesting mitigation.

Whilst these modifications have been relatively minor they have improved the overall usability of the system which helped to optimise the performance of the Operators whilst conducting their tasks. As noted previously, looking at human error at an individual component level is not sufficient in isolation when developing a complex socio-technical system such as C2. However, the Heuristics reviews provided the opportunity to identify and design out, with relatively simple software changes, potential human error by ensuring the interface aligns with the key usability Heuristics.

3.3 Assessing the system as a whole

The benefits of adopting the Complex-Systems HRA are that it enables identification of more complex error scenarios to be established which more traditional approaches would not have been able to. For example, incorrect data/information entered into the system by one Operator, may impact on another Operator's SA and therefore impact decision making. This technique allows suitable mitigation to be put in place to reduce the likelihood of these errors from occurring.

One of the benefits of adopting the approach presented within this paper is the added confidence and assurance that is provided to the overall reliability argument which is presented within the Safety Case. The proposed approach goes beyond more traditional approaches of HRA and therefore is presenting a more robust human reliability argument.

The Complex-Systems HRA approach presented within this paper is detailed and time-consuming. Therefore, it is crucial that HF efforts are focused on the areas which would benefit from this detailed approach the most. Namely any aspect of the system which is safety-related, any new aspects of the system and new functionality which is highly complex. By applying a rigorous and systematic screening exercise in Task 1 as detailed within figure 4, it will ensure a consistent approach is applied and only aspects of the system which require further detailed assessment are included.

The development of the use-cases cannot be achieved by HF in isolation. These need to be developed in consultation with SMEs and software Developers. Use cases need to vary and take account of the different operational contexts in which Operators will use the system. This is where the scenario development (as detailed within Task 3 of figure 4) is important to ensure any emergent properties are identified in each of the varying scenarios. As noted, this can be a time-consuming exercise. However, once the detailed use cases are developed, these not only serve to support the Complex-Systems HRA approach but can also have other uses in a project, such as:

- Be used by HF practitioners throughout the User Trials to ensure Operators complete tasks as expected and the system 'behaves' as expected and to identify any Operator errors; and

- To model the impact of design changes by tweaking the test cases to ensure it results in the desired outcome.

3.4 Challenges of the approach

The HRA is an on-going process specifically designed to be undertaken throughout the development phases of the C2 system to identify potential human errors which Operators may perform whilst interacting with the 'system' and. where practical, influence the design to eliminate or reduce potential human error. However, whilst the Heuristic reviews could be undertaken at early stages in the development, the more detailed 'holistic' assessment to assess human reliability requires a more mature 'system'. This is because the new functionality has to be mapped to existing tasks and roles or new procedures agreed on how the new functionality needs to be used within the existing system.

Therefore, the HRA approach needs to tread a fine line between being based on a mature enough system where it has been agreed how new functionality will be used (which goals it will support) and who (which Operators) would use it and not being too late in the development process where it is too late to influence design of the C2 system. An HRA conducted on a complex system such as C2 too early will need to make numerous assumptions about how the system will be used to achieve the task goals and Interactions within the system may not be properly understood. This could result in important emergent properties being missed. Leaving the HRA too late in the design will result in the development work being completed and implementing recommendations from the HRA work could be costly to the project.

The complex system HRA presented within this paper is ongoing and due to the stage in the project there are uncertainties in exactly how the functionality will align with current Operational procedures. However, the approach presented involves SMEs and End-Users, together with an understanding of system and user-based requirements, has allowed a number of key assumptions to be made at the right stage, which are then validated and evaluated as tasks at a later stage of the process. This is key to making an early start in the design process, and critical to the success of the HRA.

The User Trials provide a valuable opportunity to validate assumptions such as how the functionality will be used in an Operational setting and also about identify how the Operator will perform in a complex system such as C2.

4 Conclusion

The HRA approach presented within this paper highlights two key aspects of HRA. Firstly, adopting a Complex-Systems HRA approach has the benefits of identifying undesirable emergent properties, the complex interactions which occur within the system and the direct and indirect PSFs which may impact on the interactions within the system. Secondly, the integration of the HRA process throughout the development of the C2 system has the significant benefit that HF can influence the design at the early stages of the development of C2.

Not adopting a Complex System HRA approach may miss the complex interactions which exist within a system such as C2. By involving HRA throughout the development of C2 it results in design changes which can eliminate or reduce likelihood of undesirable emergent properties arising from interaction of components. This results in a system which is designed to support Operators in achieving task goals more reliably.

Acknowledgments I would like to thank Nicola Fairburn for working on the approach with me and the time we have spent discussing HRA within complex systems. I would also like to thank Frances Ackroyd, Joanne Kitchin and Kayley Clusker for their reviews of the paper.

References

Embrey, D.E. SHERPA: A systematic human error reduction and prediction approach. Presented at the International Topical Meeting on Advances in Human Factors in Nuclear Power System Knoxville Tennessee, 1986.

Field Manual, No. 6-0 Mission Command: Command and Control of Army Forces. August 2003 https://www.globalsecurity.org/military/library/policy/army/fm/6-0/chap1.htm, accessed on 04/11/2021

France, M.E. Engineering for Humans: A New Extensions to STPA. Massachusetts Institute of Technology Cambridge, MA. June 2017

France, M.E., Thomas, J. Engineering for Humans: A New Extension to System Theoretic Process Analysis, Int. Symposium on Aviation Psychology, Dayton Ohio, May 2017

Leveson, N.G. Rasmussen's Legacy: A Paradigm Change in Engineering for Safety. Applied Ergonomics, Special Issue on Reflecting on the Legacy of Jens Rasmussen, 2016

Leveson, N.G., Thomas, JP. STPA Handbook, March 2018

MOD HFI Technical Guide for Human Factors and System Safety 7-3 Technical Guide 7.3: Human Contribution to System Safety - HuFIMS - KiD (mod.uk) accessed on 11/11/2021

MOD, Joint Technical Note 2/17, Future of Command and Control. JCN 2/17, Future of Command and Control (publishing.service.gov.uk), 2017, accessed on 14/11/2021

Nielsen, J. (1994b). Usability Engineering. Boston: Academic Press.

Salmon, P.M., Walker, GH., Ladva, D., Stanton, NA. Measuring situation awareness in command and control: comparison of methods study. Conference proceedings of the 14th European Conference on Cognitive Ergonomics: invent! explore!, ECCE 2007, London UK, August 28-31, 2007

Walker, GH, Stanton, NA, Jenkins D & Young MS, Stewart, R. & Wells, L. Defence Technology Centre for Human Factors Integration (DTC HFI), 2010

Thirty years of learning by accident

Graham Braithwaite[1]

Cranfield University

Extended Abstract *The thirty years that has passed since the first SCSC Symposium coincides with the author's career in safety – starting as a student at Loughborough University studying Transport Management and Planning. Back then, the UK transport sector was still reeling from major disasters – the Herald of Free Enterprise at Zeebrugge; British Midland at Kegworth; and the Clapham Junction and Purley rail accidents. On one level, such events galvanised resolve to prevent such events from recurring, but on the other, there was also a level of risk acceptance that thankfully is no longer the case. For the aviation industry, the concern was that unless there was a radical change of approach, then, based projected traffic growth, the equivalent of one widebody aircraft would be lost each week by the end of the millennium.*

Reflecting on three decades of progress, the accident record demonstrates that a huge amount has been achieved. The role of 'systems thinking' has been highly significant, not least in helping to understand the complexities of accident causation and the myriad influences on safety performance, both good and bad. Safety science has evolved as our understanding of technical failures, human performance and organisational influences have advanced. However, milestone accidents which have taken place over that period remind us that progress has often come at the price of lives lost and property or environmental damage.

---

[1] Graham Braithwaite is Director of Transport Systems - one of the eight industry-facing themes that make up Cranfield University. Transport Systems covers seven centres: Air Transport Management; Advanced Vehicle Engineering; the National Flying Laboratory Centre; Engineering Photonics; Integrated Vehicle Health Management; the Digital Aviation Research and Technology Centre (DARTeC); and Safety & Accident Investigation. The Theme operates a number of unique facilities including a SAAB 340B+ Flying Classroom, Boeing 737 Ground Demonstrator, Accident Investigation Laboratory, the Multi User Environment for Autonomous Vehicle Innovation (MUEAVI) and an Off-Road Vehicle Dynamics Test Facility. Graham is also Professor of Safety and Accident Investigation and the academic lead for the University's £67 million Digital Aviation Research and Technology Centre (DARTeC) which became operational in 2021. He is the independent safety adviser to the British Airways Board and a non-executive member of the TUI Northern Region Airlines Safety Review Board.

© Graham Braithwaite, 2022.
Published by the Safety-Critical Systems Club. All Rights Reserved.

Throughout the period, reactive safety (or so called 'tombstone safety') has remained a vital method for learning from disasters and near-misses to prevent recurrence. The concepts of blame and liability can inhibit such learning and whilst maritime and rail transport have adopted the not-for-blame approach that aviation pioneered, the threat of civil litigation and criminal prosecution in certain jurisdictions has, in some cases, grown. The cost of failure has also risen, especially as larger aircraft such as the Airbus A380 and more complex military platforms such as the F-35 became operational.

A shift to proactive safety where events are anticipated and mitigated has, arguably been led by systems engineering and facilitated by an increased availability of safety data, especially from normal operations – such as through Flight Data Monitoring (FDM) programmes. By identifying precursor events, interventions ranging from training to advanced autonomy have helped to develop highly reliable complex systems. As data analytics continues to improve, so too the potential for predictive safety has grown.

A great deal of change in safety thinking has occurred over thirty years. At this important anniversary for the Safety Critical Systems Club, it is worthwhile reflecting on the many elements that have delivered safety performance, including the lessons learned from major events that have happened over that period. Looking specifically at aviation as an example, how has the blend of hard engineering and softer human sciences delivered gains in safety performance, especially in at atmosphere of intense commercial pressure?

Looking forward at the next 30 years, how well placed are we to embrace the opportunities presented by, for example, new materials, technologies, autonomy and artificial intelligence? If a new technological revolution based on opportunity was not enough, how does this interact with the economic effects of a global pandemic and the chronic threat presented by climate change. Can we continue to deliver superior levels of safety performance at a cost which is acceptable to the consumer, society at large and our planet?

"Cowboy[1] digital" undermines safety-critical systems

Harold Thimbleby

See Change Digital Health Fellow
Wales, UK

Abstract *A "cowboy builder" is someone who, operating beyond their level of competence who builds unsatisfactory, often unsafe, structures. This paper explores "cowboy digital" — the cowboy attitude that pervades digital leadership, management, development, and engineering. This article offers constructive suggestions, but until the cowboy digital culture is addressed effectively "safety-critical systems" will remain aspirational.*

> *"We live in a society absolutely dependent on science and technology and yet have cleverly arranged things so that almost no one understands science and technology. That's a clear prescription for disaster."* — Carl Sagan

1 Introduction

Engineering makes technology work, but much engineering is hidden from sight, and as technology matures more and more becomes concealed. Clocks were once large and their gears and pendulums were things that could be watched in action, ticking, displaying their engineering for admirers. Old steam engines. Old cars. Old corn mills. Old crystal sets. Even historic computers like Babbage's Difference Engine, or classic cryptographic machines like the German Enigma.

A modern clock has little working in it that a human eye can discern; and if it is a dependable clock, it will get its idea of the time it displays through the invisible internet and a distant mesh of atomic clocks.

In fact, hiding detail is a standard technique for managing complexity and, hence, for building safer systems.

---

[1] Cowboy: (1) a man who herds cattle, or (2), in the sense here, a person of any gender who is reckless or careless. Cowhand is a gender-neutral term for (1) but is not used in sense (2).

© Harold Thimbleby, 2022.
Published by the Safety-Critical Systems Club. All Rights Reserved

In many technologies, some or all of the concealed engineering can help ensure safety. Thus, in a modern building, careful calculations on load-bearing, fire resistance, electrical earthing, and more, result in hidden structures that ensure the building is safe. Since these structures are hidden from sight, there is a temptation to cut corners. Cowboy builders are everywhere, but — in principle — managed through health and safety regulations, buildings regulations, and more, with sanctions such as criminal prosecution. Not all cowboys are deliberately incompetent there are many unconsciously incompetent people who build unsafely; shoddy electrical and gas installations are dangerous and unfortunately common. In the "arms race" between cowboys and safety, building regulations require installations to be regularly checked (especially at legal reviews, such as when buying and selling) using formal processes by certificated competent engineers.

In comparison with building, and other physical engineering, digital technologies have developed very recently and at a dazzling pace. As video game and science fiction special effects prove, there is no connection between what we see and the engineering structures that implement the experience. Indeed, many digital systems are deceptive: they look pleasing, but all implementation details, including any cowboy implementation, are hidden. This is just like dangerous wiring in a house: the sockets may look fine, but they may hide the fact that there is no earthing, or that the current rating of the wiring is inadequate. At least anyone can pull up floorboards or unscrew the cover of a mains socket, but code is *really* hidden — it is generally somewhere else, and is sometimes lost or has mismanaged version control. Software is only very rarely available in any form for independent scrutiny.

While there is substantial regulation for electrical wiring, regulation for digital has not kept up; indeed, there are many who argue that "regulatory burden" is stifling digital innovation. It follows that digital cowboys, both unintentional-cowboy and intentional-cowboy, are common.

Cowboys of any sort create higher expectations than can be reliably delivered — cowboy salesmen promise snake oil, deluding the naïve and gullible. In conventional engineering, cowboy organisations may do this deliberately; in digital engineering, much of it is accidental as developers themselves get caught up in the excitement of pervasive over-promised digital visions.

This is obvious at the highest levels. The current various digital strategies of the Governments of England, Wales, and Scotland are full of "digital innovation," reducing regulatory constraints, and promising unsubstantiated benefits, especially with the excitement of AI.[1] Where problems are acknowledged, these are revealing: cybersecurity is acknowledged — but the cybersecurity problems *are caused by other people*; digital exclusion is acknowledged — but the exclusion problems *are people missing out on unqualified digital benefits*. In short, these strategies treat digital as an automatic solution, and newer digital as an even

---

[1] For example, https://www.gov.uk/government/publications/uk-digital-strategy

better solution. At international levels, documents such as the UN's 2030 *Agenda for Sustainable Development* says digital has "great potential to accelerate human progress, to bridge the digital divide, and to develop knowledge societies." Indeed, it may, but to do so, it needs to work reliably, which in turn relies on managing cowboy thinking.

We now know that X-rays and radium are dangerous because radiation can cause cancer, yet at the time they were discovered they were celebrated because of their dramatic properties. Their soon-to-be-discovered problems were invisible and unrecognised. "My beautiful radium … What can it *not* do?" asked Marie Curie, winner of the Nobel Prize in chemistry for discovering it,[2] and later herself to die of cancer caused by it and her other radiation research (Moore, 2016).

Digital computers are not radiation of course, but they share the same properties that Marie Curie highlighted: invisible bugs and unrecognised, uncontrolled problems, and a huge popular excitement about them. People then did not know what they didn't know, and it turned out to really matter. Von Sochocky could have been talking about digital when he said, "Locked up in radium is the greatest force the world knows … invisible forces the uses of which we do not yet understand."

It is ironic that one of the landmark events in the field of software engineering was the Therac-25 (Leveson, 1995): a computer-controlled radiation treatment system that, because of bugs, killed patients.

There was a period when radiation was not regulated (and when manufacturers of luminous and other clever products would have opposed regulation), but radiation's novelty and new behaviour was not an obstacle to effective regulation and carefully controlled processes for using it. This paper argues that the unrecognised and hidden complexities of digital lead to cowboy behaviour, analogous to selling glowing snake oil. The safety issues digital is increasingly raising should encourage us to improve quality controls and regulation so safety is improved.

2 Safety is a privative

"Safety" is a *privative*: it is the avoidance of unwanted harms, or, more generally, reducing the risk of harms to an acceptably low level. The concept of "privative" (due to Aristotle) is critical: it means that safety is an *absence* of things. The word safety creates the impression that there is a concrete *thing* that can be called safety, but safety is not a thing. Safety is the absence of unwanted things.

---

[2] Marie Curie and her husband Pierre won the Nobel Prize in physics (shared with Becquerel) for their work in radiation; then, after his death, Marie won the Nobel Prize in chemistry for discovering radium and polonium. She was the first person to win the Noble Prize twice. She is the only person to win the Nobel Prize in two different scientific fields. The Curies' laboratory notebooks are still too radioactive to be safe to touch.

One consequence of safety being a privative is that it is not possible to directly measure it. Instead, we measure harms (i.e., defects or prevented harms) and assume zero measured problems implies safety (or that fewer mean improved safety, which is also fallacious). Unfortunately, no measured harms means only that we have not yet encountered some of the possible future harms that may occur: it does not mean a system is safe; at best it means that the system was found to be safe *during testing along the specific paths where it was tested* — and even that is assuming the tests themselves are correct and can correctly identify faults. Yet the main point of measuring safety is to estimate the risk of *future* harms![3]

One might think that testing is a good way to establish a program is running correctly, and is safe. However, as famously pointed out by Edsger Dijkstra, testing can only find out whether things are right or wrong at most where it is tested; testing cannot prove the absence of defects anywhere else. In other words, testing can never show a system is safe. Instead, although testing can help find problems, fundamentally, software has to be designed proactively following appropriate engineering standards to be safe (or as safe as reasonably practical). A more mature view is that if testing finds problems, then the assurance processes used in building the system failed and *they* need fixing too, not just the specific defects identified through testing.

Here's a small concrete example. On the iPhone (version 14.8) Apple calculator, the following sequence of keystrokes results in an error:

AC 5 ± [swipe finger across number to delete] =

This bug can be fixed by pattern-matching that sequence of keystrokes and displaying the right answer instead of 'Error'. However, the error is more general than the test reveals: anywhere during a calculation, entering any number > 0 or more than one 0, ± [swipe to get down to zero] = will result in an error. In other words, the failure of the simple test is not so much a single bug to fix, but an indication that the mappings between the user interface, the numbers displayed, and the calculator's number representations are faulty, and more generally as I have reported this type of bug to Apple for years, that their quality control (or their desire to have a reliable iPhone calculator) need fixing. Therefore, there are likely many more bugs. Given that Thimbleby and Cairns (2017) point out that such problems are not unique to this specific calculator, one concludes that current handheld calculators are not dependable and should not be used in any safety-critical application *even if* specific bugs can in principle be avoided. They are

---

[3] There are many more sophisticated methods for testing. For example, randomising test data, or fault injection which inserts known defects, then a sample of independent tests is used to statistically improve the estimate of original defects. Unfortunately, bugs that may cause harm need not have the same properties or distribution as (typically trivial) inserted faults.

cowboy digital: they look good, but their flaws are hidden except to careful inspection.

A problem with cowboy developers is they do not have the memory and generalised thinking of professionally-trained hindsight. They may think their work is successful, but they are unaware that their testing may be inadequate and that their apparently successful work to date gives them no assurance whether their next project will be successful.

They are unaware of standards, processes, and defensive methods, and often unaware of key stories (e.g., the Therac-25). Their personal experience may be no more helpful than "nobody has had problems with my software, therefore it is safe." With proper training, they would know their personal experience is success biased and therefore inadequate. Without rigorous qualifications that assess software engineering safety skills their employers cannot know their inadequacy either.

In physical engineering, it is routine to test structures at much higher stresses than they would normally operate at. Consider RCDs (residual current devices),[4] which are electrical safety devices, and serve an analogous role to safety railings on balconies — most of the time they are not needed, but when they are needed they must work reliably. Since RCDs are safety devices, it is important to ensure that they work correctly. The RCD tester, explained below (shown in Figures 1 and 2), for example, tests RCDs at 5 times their rated current. The idea is that "more" means a better test — if something critical works properly at 5 times some rating, then it will likely be safe at its design rating. In contrast, in software, in general we have no idea how to do "more" so easily — indeed, unless it is very carefully designed, it is hard to identify modules like "RCDs" that can usefully be tested independently, and which can be stress tested directly (Rodriguez, et al, 2014).

Software's relation to safety adds another complexity. Software alone, by itself, does nothing, so it can neither be safe nor unsafe — software instructs computers how to do things, which *then* may be safe or unsafe in various *contexts*. In this sense, software is more like engineering calculations: the calculations may be just squiggles on paper, but they specify how structures are built and work. However, turning an engineering calculation into a specified artifact requires skilled work. Software, in contrast, has no such step; once it is on a computer, it works on its own and requires no human intermediaries. There is no skill threshold between program and process.

---

[4] RCDs are called ground fault interrupters, GFIs, in the US.

3 Illustrative examples

Cowboy digital problems happen, happen often, and happen on a large scale. The ease with which cowboy programming appears to work has led many programmers unwittingly to work beyond their abilities, often with neither them nor their employers aware.

The following examples are chosen because they can be briefly explained — one of the serious problems with addressing cowboy digital is that bad systems and bugs are often tedious to explain in sufficient detail to be reproducible, and people lose interest in the details. I hope then that these few examples are interesting as well as informative.

The examples below show how elementary software defects are routinely ignored — and they are not just ignored, they are rarely even noticed. Our social culture is that this is just how it is, so the usual solution offered is to buy a new one, but without ever explaining how the new one will avoid the problems of the old. This is one of the flaws in the UK digital strategies discussed earlier: they are keen to emphasise that digital is now newer, as if that alone means it is better. National digital policies should not be treated like upgrading consumer products.

This first example below has been chosen because it is in an area that is tightly regulated, with wide-ranging scientifically-literate standards that have legal status. Furthermore, the example is a current product, which is widely available on the market. The accuracy of my discussion is easily checked.

3.1 Software-concealed "safety" features

This example is about managing electrocution, a safety concern that we all recognise, and one where safety features are often implemented by software.

In the UK, domestic electricity is provided over line[5] and neutral wires, along with some form of protective connection to earth. If the current in these wires exceeds some limit, such as 20 amps, a circuit breaker will trip and isolate the faulty circuit. Unfortunately, humans would be electrocuted at such levels of current, so, while circuit breakers help ensure safety (helping avoid fires), they do not protect against electrocution.

In contrast, RCDs measure the *difference* between line and neutral currents, which will be very close to zero in normal use, even when powering a heavy load like a heater. Any difference in line and neutral currents is a fault current going

---

[5] Previously *line* was called *live*. The change in terminology reflects that under many normal conditions or faults *any* wire may become live. The word *live* now means energised or at risk of being energised.

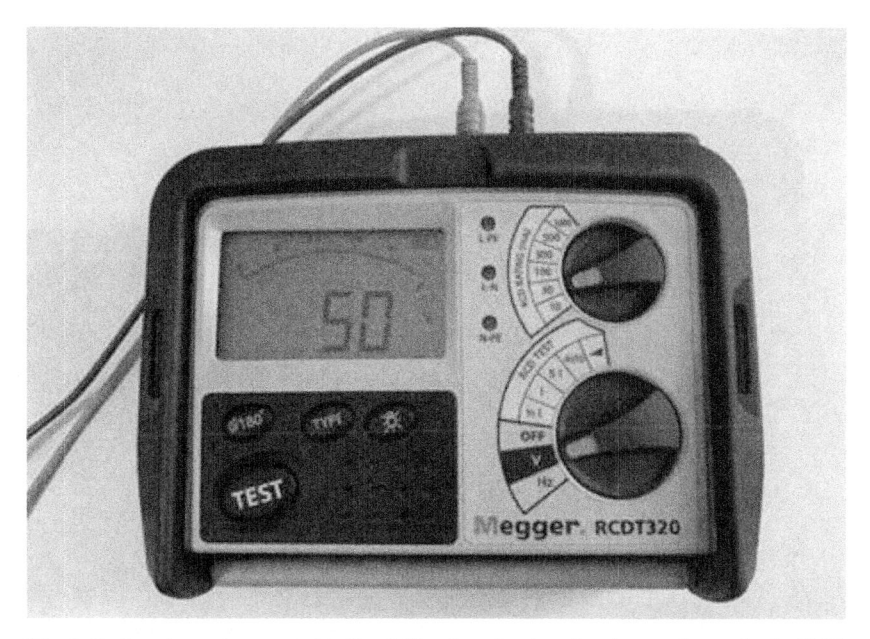

Fig. 1. Test equipment connected to live 245-volt mains, but showing a voltage of only 50 V. Confusingly, the voltage shown is the instrument's touch voltage limit, which is nothing to do with the voltage across the leads. Compare this with figure 2.

to earth, for instance through a human body standing on the ground. Common RCDs with fault currents of only 0.03 amps (30 mA) will trip in milliseconds, so they protect people against electrocution yet allow large currents flowing equally in line and neutral wires to operate high-current appliances like kettles, heaters, or cookers.

RCDs must work reliably, so there are devices that test RCDs (including the RCD's basic test buttons).

Figures 1 and 2 show a popular RCD tester, the Megger RCDT320. RCDs have to be tested connected to the mains, under live conditions. Live working can be dangerous, and equipment designed for live testing therefore has to be designed to be safe. Equipment designed for testing electrical devices must therefore stop tests if voltages the user is exposed to (so-called touch voltages) exceed safe limits, such as 50 volts AC, or lower (e.g., 25 V AC) in wet conditions.[6]

The Megger tester conforms to standard touch voltage requirements, yet it has no visible features for setting or confirming its actual touch voltage settings. The developers introduced program code that sets and checks touch voltage limits

---

[6] The higher the touch voltage, the higher the current that would pass through a body to earth. A high-enough current for a sufficient period of time will cause the heart to arrest, although much lower currents can cause a person to jump, and perhaps fall and injure themselves.

210   Harold Thimbleby

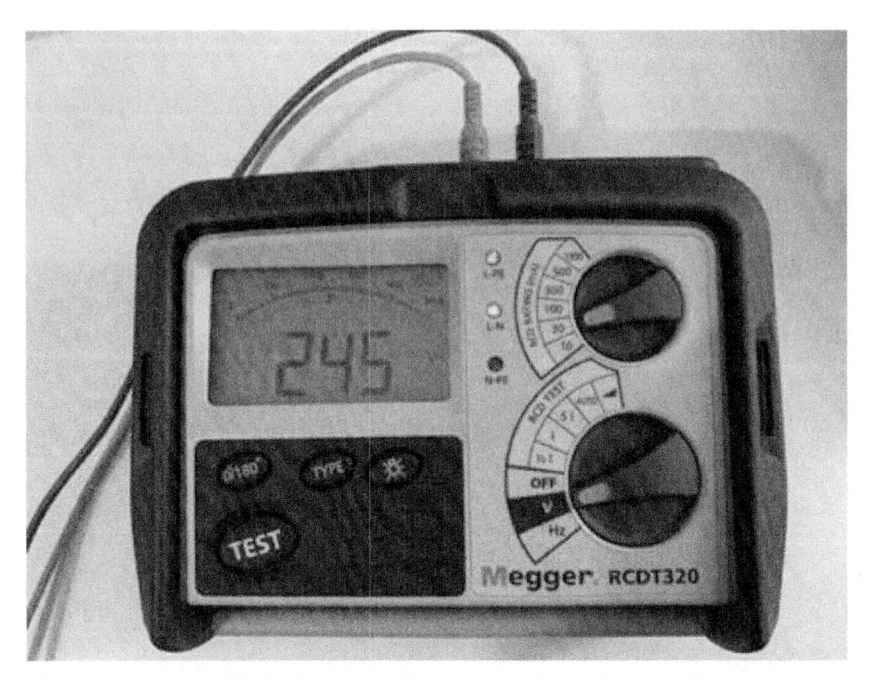

Fig. 2. The same test equipment connected under *exactly* the same conditions as shown in Figure 1, with all buttons and knobs in the same configuration. The voltage displayed is now correctly shown as 245 volts. Two of the three live warning indicator LEDs are also illuminated (L-PE and L-N): it is a bug they were not warning of any live connection in figure 1. Clearly, the warning lights are programmed as *guesses*, as with only two connections, as here, the tester can only *know* one of the three cases for certain; here, it happens to be connected to L-N only, so the L-PE indicator on and the N-PE off are just guesses. (It is safer to guess L-PE is live than not, of course, but if safety was a serious consideration, it ought to have lit N-PE too, as it cannot be certain of the status of that either.)

invisibly. It is as if the Megger design was completed, and then new touch voltage software features were added.

Here is how the user changes the touch voltage limit, lightly edited from the user manual for clarity:

> With the instrument switched OFF, hold down the [TEST] button. Turn the range knob to any ON position. Keep the button held down until the instrument displays 'SEt' (on a 7 segment display). Now release the [TEST] button. Press the [TEST] button twice. The display now shows '25 V' or '50 V'. Press the [0/180] button to change the limit from 25V to 50V and back. Press the [TEST] button to exit from the set-up menu.

It is notable that none of this process has any affordances (visible cues) in the physical design of the tester. There are no cues visible in Figures 1 and 2, and none are presented on the tester's summary instructions on its lid.

Figure 1 shows the tester with the knobs set to measuring voltage, with the device connected to 245 V mains, but misleadingly the tester is displaying only

50 V! The tester is displaying the touch voltage it is set to, not the voltage it appears to be measuring. This is very dangerous.

Figure 2 shows the same tester connected under *exactly* the same conditions to the mains, but it is now showing the actual voltage the tester is connected to, as well as lighting two red indicators confirming that the wiring is live, specifically warning of live L-PE and L-N voltages. These indicators are helpful, but in fact they are misleading guesses — the tester only has two wires (as shown in the figures 1 and 2) so it can only *know* that one pair of wires is live. As shown, it happens to be connected to L and PE, and the tester knows nothing at all of any L-N voltage.

During setting the test voltage limit, these indicators were unsafely and incorrectly *not* illuminated (Figure 1). This is a symptom of cowboy programming: a feature (here, touch voltage interaction) has been added to a program apparently without understanding what safety properties the program should have, and has actually introduced some new unsafe bugs because, it would appear, the interactive feature runs but does not complete before the program has got far enough to illuminate any warning indicators.

In short, on this device, there is a software safety-critical feature that is invisible, obscure to use, unmemorable, and buggy. It seems plausible that the documentation writers, who tell users how to operate the device safely, were unaware of details of the device's programming (or of its significance): the tester has a prompt sheet on its lid, but provides no prompts for these features, nor does it warn of dangerous ambiguities in voltage displays.

This paper's appendix shows that, additionally, the tester's programmed measurement of touch voltage itself is buggy, which could cause more unsafe use.

3.2 The Post Office Horizon scandal

The previous example is not undermining Megger, the company that makes the device, as it was presented as a representative example that can be checked independently. Many more similar examples could have been given in a longer paper!

The UK Post Office pursued nearly a thousand prosecutions of people it accused of theft or fraud (Wallis, 2021). To do so it relied on computer evidence from a system called Horizon. It is now public knowledge that Horizon was buggy, and that, as the Court of Appeal put it, the Post Office subverted the integrity of the criminal justice system and public confidence in it by pursuing prosecutions that claimed otherwise. Its prosecutions were declared an abuse of process and an affront to the conscience of the court. Arguably, it is the largest miscarriage of justice in the UK ever, resulting in numerous unwarranted prison sentences, loss of homes, divorces, and suicides. Clearly, the failure of "merely" mission critical software causes significant safety failures. Tragic is not a strong enough word.

A professional system such as Horizon can only have unnoticed and uncorrected bugs of such significance, such as accounting errors, if it is cowboy programmed. More specifically: the failure to check whether alleged accounting errors, once they were noticed, were bugs or human accounting errors (and therefore possibly fraudulent) was negligent, and certainly a failure of following normal auditing procedures. Furthermore, the failure to design the program to adequately self-check was negligent (or, if it did adequately self-check, to ignore those checks), especially considering that accounting has a long history of error checking processes.

Plausibly, programmers were not competent or not motivated to test thoroughly or to insert code that would detect their own programming errors. It is plausible that, at least at first, programmers did not disclose to management that Horizon was buggy, perhaps because to do so would have put their jobs in jeopardy. It is likely, then, that management was in thrall to a "complementary cowboy effect": possibly, they believed a buggy system was not buggy because they knew no better. The Post Office then pursued a vast number of prosecutions, showing no curiosity (and, later, deliberate denial) why there were so many problems.

Horizon going wrong in the first place was cowboy programming. Subsequently, a core issue in the subsequent failure of justice was the legal presumption that computer evidence is correct. This inexorably led to evidence — misleading evidence based on buggy software that was not rigorously examined — being withheld from scrutiny by the courts and the defendants, because it was presumed correct. This is cowboy digital thinking affecting the law (Ladkin *et al*, 2020; Marshall *et al*, 2021).

3.3 The ECM Synchronika espresso machine

The ECM Synchronika is widely recognised as a beautifully, well-engineered espresso machine. In other words, ECM have skilled engineers and know how to use them to make well-engineered, substantial products. Many dealers selling the Synchronika provide photographs and videos on their websites of its internal engineering: compared to other espresso machines, its engineering is a work of art.

The Synchronika is not so much a safety-critical example as an example of the ease with which skilled conventional engineers regularly ignore competent software engineering.

A good espresso machine must heat water to a precise temperature, such as 93 C, or whatever is preferred by the barista as appropriate for the brew. The

Synchronika uses a PID, a computerised temperature controller that uses proportional, integral and derivative (i.e., PID) feedback to control the water heater. On the Synchronika, the simple-looking PID user interface has a three digit, seven segment LED display and two buttons. It is impossible to use its many features without reference to the user manual because the three digits give far too little feedback.

The Synchronika can be plumbed into a water supply, or it can use an internal water tank. If so, it is of course important the user knows when the tank is empty. A cowboy programmer feature on the Synchronika shows the condition of no water: completely blank out the LED. If the LED is lit, showing anything, the tank has water; if it displays nothing, the tank is empty — or, ambiguously, the machine is switched off.

About once a month, somebody in my house comes to me and says "The coffee machine is broken! It's not working!" But it *is* "working" it is just that the programmers decided to show the empty tank state by making the machine apparently show it is not working.

The irony is, no expense was spared on the visible conventional physical engineering, but corners were cut in the software and in the user interface design.[7] The implication is that the manufacturers (their employees) are unaware of cowboy software engineering skills despite being very aware of conventional engineering skills.

If no additional investment was available for a better display, the programmers could have made the existing LED display flash "H2O," say, which would have worked on the seven-segment display. A more satisfactory solution would have included a display that had adequate resolution to display the machine's status and stages in all user interface interaction clearly.

Here's a conversation reported in Brian Kernighan's *UNIX: A history and a Memoir* (Kindle Direct, 2021) illustrating a Unix Systems Laboratories (USL) manager's software skills:

> **Manager**: "You have to fix all the bugs in the C++ compiler, but you can't change the behavior in any way."
>
> **Brian**: "That's not possible. By definition, if you fix a bug, the behavior is necessarily different."
>
> **Manager**: "Brian, you don't understand. You have to fix the bugs but the compiler's behavior can't change."

---

[7] Another bug is that if the tank runs empty when making coffee, the user will of course try to lift the tank out to refill it. Lifting the tank raises the height of the depth float, so the pump comes on again to continue pumping the water that it now buggily thinks is in the tank. This makes the tank impossible to remove without switching the machine off to kill the pump suction.

3.4 Further examples …

The daily news is regularly supplied with examples. Here's just one: after having spent nearly £800,000 on cyber-resilience, the Scottish Environment Protection Agency (SEPA) had over 4,000 files stolen in a cyberattack. SEPA's response plan was inaccessible during the incident as it was on the servers that were attacked — and there were no other versions (BBC, 2021).

The Safety-Critical Systems Club has a regularly-updated web page collecting details of further examples of cowboy programming.[8]

My own book *Fix IT* (Thimbleby, 2021a) documents diverse safety-critical examples from healthcare, including a "trivial" Y2K cowboy bug that killed people, and the investigation of which did not understand digital either — the cowboy digital extended upwards to official levels. Thimbleby (2020 & 2021b) cover further examples that appeared after the book went to press, and the classic Leveson (1995) is also recommended.

AI provides regular examples, notably in failures of driverless cars. From a software engineering perspective, AI/ML is just complex software, some of which is parameterised, effectively written, by the AI itself as it learns. AI is now widely used in areas that obviously have direct social impact, and society seems to be realising that AI needs ethical oversight and perhaps further regulation. This is because many bugs from cowboy AI programming become visible in decisions (e.g., credit worthiness, access to healthcare, sentencing) that non-technical ethicists recognise as raising clear safety and ethical issues (including fairness, reciprocity, racial bias, privacy, poverty). Good discussions can be found in (Garvie *et al*, 2016) and (Hao, 2020).

4 What to do?

Compared to house building, an engineering example already mentioned, software engineering might be considered analogous more specifically to structural engineering, and (to that extent) coding analogous to bricklaying. Bricklaying is an essential skill to construct almost any building, but however good or skilful bricklaying may be, it needs to be closely guided by professional structural engineering to build anything sophisticated that is safe. A bricklayer might be able to make a nice garage for your car, but they would be over-stretched to make a tower block, as it would require sophisticated calculations to ensure its safety.

Society recognises that many technologies must be adequately safe. For example, there are regulations to ensure that competent electrical engineers undertake domestic electrical wiring, work to strict standards, have their wiring

---

[8] See https://scsc.uk/f138 or search for "Safety in the news" within that site.

checked, and who may face criminal sanctions if their work is substandard. No such stringent regulations specifically apply to digital systems engineering — even though digital systems now often control electrical systems!

4.1 A diagnosis

The diagnosis of the various problems discussed in the examples above, and others in the wider world, is that (*a*) professional software engineering as such does not have a sufficiently prominent role in design and development; (*b*) software engineering does not have a prominent place in regulatory and legal frameworks; (*c*) software engineering does not have a rigorous qualification and certification process enforced for (*a*) and (*b*). Furthermore, cowboy programming is apparently so successful, there is no pressure to fix points *a*, *b*, *c*.

In addition, software engineering already has many devalued qualifications. Qualifications like CITP[9] or SFIA,[10] which reflect the needs of business, primarily for operating and managing IT systems — not in competent software engineering or implementation. These qualifications are analogous to electrical qualifications required for wiring houses — such qualifications are important, they may well be challenging to get, but they do not say anything about skills for designing new circuits. Almost all software engineering is developing new computer systems, and considerably higher-level skills than are required for using and managing completed systems are required. The qualifications we will talk about below are envisaged to be higher-level skills, post-graduate certainly, and often post-doctoral. Note that the innovation in "digital innovation" relies on pushing the boundaries of what computers can do, which is research by any other name, and it needs treating by competent software engineers with research skills. The diagnosis, then, of cowboy digital is that it isn't developed by people with adequate competence (despite their basic qualifications); that is, novel systems are developed as if development is easy, which is exactly how cowboy developers of all sorts behave.

4.2 High-level solutions — to prompt debate

Society must recognise that many digital systems are safety-critical. However, merely tightening regulation is not sufficient, not least because the usual suspects who write regulations are subject to the same social forces that manifests itself in

---

[9] http://www.bcs.org/media/1062/chartered-it-professional-standard.pdf

[10] http://sfia-online.org/en

cowboy digital. Even with regulations, someone or an organisation unaware of what they do not know ("unconscious incompetence") can perform risk analysis and document risk analyses, etc, to conform, so far as they know, to any regulations. Just as good programming requires independent code review, adhering to regulations for software requires independent inspection and review.

Therefore, regulation must be matched with improved education, continuous professional development (CPD), and accreditation, along with "licence to operate" processes to ensure that enough developers are professionally competent for the types of programming (supervision, risk analysis, etc) that they undertake. Regulation has to be developed and written specifically to enforce software engineering concerns. Despite even substantial scientifically-literate regulations like BS 7671 (*Requirements for Electrical Installations*), etc — as Example 1 highlighted — regulation without software insights is ineffective against cowboy digital. It should be noted that some caution is required with considering education as a solution (e.g., Fisher & Keil, 2016); it is not a panacea, as it can lead to counter-productive over-confidence in one's competence.

Effective digital regulation will not happen overnight, but will require long-term investment and commitment, including deliberate iterative development, as well as a change in culture that recognises cowboy programming as an avoidable risk. Until it happens, many critical professions will continue to use incompetently regulated digital systems that undermine their own professional safety standards. For example, an anaesthetist rightly has years of training, substantial qualifications, professional accreditation, continuous assessment, and more — yet when they press a button on an anaesthetic machine the effect on the patient may be determined by a cowboy programmer (Thimbleby, 2021a). There is a virtuous circle: just as improved regulation will improve education (e.g., requiring certification, etc.) so improved education will, when the (post)graduates get on to the right committees, improve regulation.

All regulations cited in this paper are expensive (e.g., BS 61010-1 is £350); the BS 7671 standard refers to 150 other standards, which themselves refer to further standards … and so on. "Effective" regulation needs to be much cheaper than standards currently are, otherwise developers are not going to have ready access to them. Safety regulations should be subsidised: if they are read and used, they make a public health saving.

Robust test systems (the analogue of electrical regulation-conformance testers, such as RCD testers) need to be developed that are capable of assessing and helping audit digital systems against the relevant standards. Such systems would themselves be covered by relevant regulations to ensure reproducibility. The Horizon case emphasised the need for error logs to be made available, and, conversely, failure to make such necessary auditing data available ought to discredit the evidence.

There are many more perspectives and approaches. Indeed, software engineering has many opportunities to exploit automation to make it more effective that are not available to electrical systems, which formed the basis of the analogy above. Model checkers and theorem provers, type checking, and other rigorous systems, are readily available.

This is no easy call, but until we start, we will never have dependable safety-critical systems in widespread use, let alone where they are needed. Society will repeatedly fall for the siren cowboy calls that now is the time to develop new, exciting digital systems to fix the problems we currently have, yet with no learning taken forward from the previous failures.

4.3 Rating programmers

Unconscious incompetence results in over-confidence and people attempting tasks beyond their abilities. Furthermore, such people tend to confidently over-rate their own abilities because they cannot see their own mistakes; conversely, people who are good programmers may be acutely aware of their limitations, so more software ends up being developed by incompetent programmers. Such problems are universal, and impact all skilled work (Dunning and Kruger, 1999).

While there remains no way to objectively assess programmers, it follows that the wrong people will dominate digital leadership and management, as well as dominating development itself. The visions and strategies put forward for digital innovation will be unrealistic, and the developers implementing those visions will not even realise.

In many areas of life, competence is typically assessed by examinations (such as degree qualifications) and competitions (such as races). Chess has several systems for rating player competence. The Elo rating system is simple; basically, rated players play games, and after each game the players' ratings are adjusted by formula. From the resulting Elo ratings of two competing players, one can estimate the probability of each winning. The rating is a good estimate of a player's estimated ability to understand positions and improve them to win.

Top performers in an intellectual domain like chess beat experienced amateurs as easily as an amateur can outperform a beginner — that is, almost all the time. Moreover, a *blindfolded* chess expert can beat several amateurs playing at once, whereas an amateur might make mistakes and would have to work to beat a beginner. The skills of top players seem like magic, just as the skills of beginners are as clumsy as cowboys. Could we have more software competitions and introduce an analogue of Elo ratings?

Programmers can be assessed more directly, say, by counting lines of code divided by bug counts. Clearly, some assessment could be automated, and some assessment could be provided to programmers directly from their production tools. Developers would then be aware of their skills, and employers could competitively select developers of sufficient competence to implement safety-critical systems. Salaries would reflect competence, and the employment market would encourage programmers to improve their skills, or — if the rating system separately assessed skills in graphics, finance, user interface, machine learning, security, safety, etc — better match their skills to specific types of programming.

Of course, some organisations already use ratings. Use of the Capability Maturity Model Integration (CMMI) is required in many US Government software contracts. While the CMMI was developed by the Carnegie Mellon University's Software Engineering Institute specifically for assessing software engineering maturity, there are many maturity models of varying relevance.[11] Almost all maturity models reference an organisation's maturity, rather an individual software engineer's maturity.

Note that one should take care using appropriate maturity models; for instance, the UK NHS's Digital Maturity Assessment[12] assesses how paper-free NHS organisations have or are planning to become. It has nothing to do with software engineering concerns: competencies in designing, developing, procuring, maintaining, or even in using or assessing dependable digital systems.

4.4 Transparency and open systems

Code review helps ensure more dependable systems, and open source software opens up code for review and further development potentially by any number of contributors. While open source methods can easily be used within an organisation, open source is hard to control when external access is permitted. This suggests that software engineering credentials need to be checked and enforced at least as rigorously, as, for instance, financial credit is — that is, by using trusted third parties. Of course, a precondition to do that is to have suitably rigorous professional credentials being available for rating developers in the first place.

Open source techniques have not yet been seriously used for regulatory purposes. For example, the hazard and risk assessments required by IEC 61508 are generally proprietary, and therefore no open source methods are used to help spot and fix "unknown unknowns" that an organisation working alone may be unconsciously blind to.

Open source is a special case of transparency that could be used to help improve the "going through the motions" limitations of current regulatory conformance. In particular, open source can enable academic researchers review code: software engineering research is continually pushing the boundaries, and finding new insights to improve safety. For example, Thimbleby and Cairns (2017) give

---

[11] http://en.wikipedia.org/wiki/Capability_Maturity_Model

[12] http://www.england.nhs.uk/digitaltechnology/connecteddigitalsystems/maturity-index

examples of ubiquitous safety-critical bugs in proprietary interactive systems (affecting many medical devices, amongst others) that have been very widely overlooked, presumably because programmers, even with the help of formal risk analysis, were unaware of the issues the research uncovered.

4.5 Legal responsibility

Software warranties and end user licence agreements (typically consented to before software can even be used) deny or greatly limit manufacturers' and developers' liabilities in the case of defects or harms. Examples are given in Thimbleby (2021a; 1990).

Until software warranties reach the basic moral standards of typical physical warranties, there will be little incentive for developers to take responsibility for the quality of their software skills.

4.6 Incentives

It is notable that a web site like Amazon's works very well, doubtless because market incentives help improve it. When there is a bug or when a user makes a mistake on Amazon's web site, Amazon generally has costs to fix the problem; conversely, when a user is successful, Amazon has income.

In contrast, the dire web site to access my GP (general practitioner doctor) has reverse incentives: if I fail to work out how to use it correctly, the surgery gets *less* work. In fact, my GP surgery has several web sites, which are not co-ordinated. Several times I've been lost on the "wrong one" which has no idea about the others, let alone any telephone numbers so you can escape web limitations and go back to telephone queuing. (One website says "phone your GP," but gives no number, yet in principle it knows the GP's phone number.)

There are no incentives for the developers to do the trivial work to make the web sites more helpful to use. There is nothing visible in the site for users to help the developers gain insight into its use or do continual improvement (for which there is an international standard, ISO 9241, which one might uncharitably assume the developers and managers are unaware of).

I imagine the developers "finished" the sites, demonstrated their features to managers who wanted them released immediately since the demos seemed to work.

The question, more generally, is how do we incentivise safe, dependable software? One answer would be to require CPD or other accreditation to require placements in successful digital employers like Amazon that know how to make dependable systems — just as you would not qualify an anaesthetist who had not done an adequate amount of supervised hands-on work first. Another answer would be to develop proxies for the market incentives like Amazon benefit directly from: for instance, systems could be star-rated, so more stars mean more prestige. Michelin stars seem to work; why not ACM/BCS/IEEE/IET stars? Oscars seem to work; why not galas for great software engineering?

4.7 Clear thinking …

In the physical world, which we have grown familiar with since birth, things get old and wear out. Things go rusty. Mould grows. Woodworm eats. We know that old things eventually need replacing. This reality plays into cowboy digital, because just as old things need replacing so old computer systems obviously need upgrading. Yet digital systems do not degrade over time by themselves; instead, upgrading and debugging digital systems eventually introduces so many bugs and incompatibilities that systems are made inadequate. At that point, the call to replace digital systems with newer systems to fix them then seems irresistible. Yet this is a cowboy fallacy: the reason the systems need fixing is because of bugs ("software rot"), not because of age. Newness does not fix bugs so much as change the bugs.

4.8 Campaigning

It is one thing to talk about a problem, as this paper has, but another to solve it. Cowboy digital is rife because it clusters nice, mutually reinforcing stories: the excitement of consumerism; the exhilaration of business innovation; unrestrained growth of markets. In contrast, safety problems are recognised for a day in the news when there has been a tragedy, but few trace it back to the root causes and strategies for reducing similar problems in the future.

It is fascinating to know how to make safer systems (and there is an enormous literature on it, a lot of which the Safety-Critical Systems Club has contributed) but to ensure those ideas are used in the wider world is a matter of campaigning, to get the message alive to a wider public who can change practice.

How effective is campaigning and writing reports? Well, politicians have announced £4M funding for teaching Latin in schools, because, while Latin was part of their hard education, digital consumer products like iPhones seem easy to make, easy to use, transformative, and magic, so they think there is nothing to learn there. They never learned that building safer systems requires harder work than learning school Latin. We need to change that.

> *"But the subject [Latin] can bring so many benefits to young people, so I want to put an end to that divide. There should be no difference in what pupils learn at state schools and independent schools"* Gavin Williamson, then UK Secretary of State for Education

Williamson had previously presided over a computerised national grading fiasco that gave higher grades to students from private schools and affluent areas, leaving high-achievers from state-schools disproportionately affected. Because cowboy digital has no public profile, he had not made the connection between his promoting a defective computer system and his having had an education better suited to the first century than the twenty first.

As I understand it, Williamson was motivated by a policy report (EEF, 2021) that showed that young children's language skills suffered in the pandemic. (If so, my policy would have been to boost teaching the language skills that had actually suffered and which are associated with social detriment, which would primarily be English.) But the point is: where are the safety community's influential policy reports? Who will step forward to start drafting them?

5 Conclusions

> *"You've got to be very careful if you don't know where you are going, because you might not get there."* Yogi Berra

Software is a relatively new invention, and it has driven some amazing developments in society. It is the foundation of a wide range of popular systems with which many people are familiar, from social media to spreadsheets. Software is the driver of much consumerism: indeed, we all want the latest systems and innovation. Yet digital enthusiasm, even being well-informed about the market and new products and ideas, is not the same as professional competence to understand, specify, develop, procure, operate, manage or assure safety-critical digital systems.

The examples in this paper illustrated the chasm. Somehow people and manufacturers have rush into developing and distributing safety-critical systems without sufficient regard for professional software engineering. People seem to think that software development is easy — "even children can program" — so there is inadequate reflection on how to program safely or securely. Many COVID-19 applications were apparently put together in a rush without any useful software engineering oversight (Thimbleby, 2020).

This paper argued and showed evidence that argues we need to develop processes and tools that can help assure safety, and then develop regulations that require such processes to be used by certified competent engineers.

To do that, we need to fully address the social culture — the politicians, leaders, and management, and even programmers themselves — who assume programming is easy. While anyone celebrates that children can program, nobody will require higher standards, and cowboy digital will continue.

This paper, though, is not finished. The initial list of ideas show that some ideas may help manage cowboy digital. There are surely more ideas, more ways of thinking clearly, and many which will be better than those offered here.

Enthusiastic children can do wiring, but we do not dream young enthusiasts should wire up a house. Equally obvious, we do not have national strategies for more electric cars that want to reduce the burden of electrical regulations. If we can think clearly about electricity, why do we not think as clearly with software, which can do far more damage than merely electrocuting a household?

Digital regulation is currently in the same "wild west" gap as between the discovery of radium and the discovery and enforcement of the principles behind the safe use of radiation. If we do not start gently tightening digital education and regulation now, we will enable those who wish to reduce regulatory burden to continue to build more cowboy systems people grow to rely on, and which will seem to justify delaying regulation further.

Acknowledgements I appreciate very helpful comments from Peter Ladkin, Mike Parsons, and Martyn Thomas.

References

BBC, "Hackers had second go at Sepa during cyber attack," www.bbc.co.uk/news/uk-scotland-59054590, 28 October 2021.

EEF, Education Endowment Foundation, press release, 2021. https://educationendowmentfoundation.org.uk/news/eef-publishes-new-research-on-the-impact-of-covid-19-partial-school-closures

Fisher, M. & Keil, F. C. "The Curse of Expertise: When More Knowledge Leads to Miscalibrated Explanatory Insight," *Cognitive Science*, **40**:1251–1269, 2016. DOI 10.1111/cogs.12280

Garvie, C., Bedoya, A. & Frankle, J. "The perpetual line-up: Unregulated police face recognition in America," Georgetown Law Center on Privacy and Technology, 2016. https://www.perpetuallineup.org

Hao, K., "The coming war on the hidden algorithms that trap people in poverty," *MIT Technology Review*, December 4, 2020. https://www.technologyreview.com/2020/12/04/1013068/algorithms-create-a-poverty-trap-lawyers-fight-back

Kahneman, D. *Thinking, Fast and Slow*, Penguin, 2012.

Kruger, J. & Dunning, D. "Unskilled and Unaware of It: How Difficulties in Recognizing One's Own Incompetence Lead to Inflated Self-Assessments," *Journal of Personality and Social Psychology*, **77**(6):1121–1134, 1999. DOI 10.1037/0022-3514.77.6.1121

Ladkin, P. B., Littlewood, B., Thomas, M. & Thimbleby, H. "The Law Commission presumption concerning the dependability of computer evidence," *Digital Evidence and Electronic Signature Law Review*, **17**, 2020. DOI 10.14296/deeslr.v17i0.5143

Leveson, N. G. *Safeware: System Safety and Computers*, Addison-Wesley, 1995.

Marshall, P., Christie, J., Ladkin, P.B., Littlewood, B., Mason, S., Newby, M., Rogers, J., Thimbleby, H., & Thomas M. "Recommendations for the probity of computer evidence," *Digital Evidence and Electronic Signature Law Review*, **18**:18–26, 2021. DOI 10.14296/deeslr.v18i0.5240, https://journals.sas.ac.uk/deeslr/article/view/5240

Moore, K. *The Radium Girls*, Simon & Schuster, 2016.

Rodriguez, I., Llana, L., & Rabanal, P. "A General Testability Theory: Classes, Properties, Complexity, and Testing Reductions," *IEEE Transactions on Software Engineering*, **40**(9):862–894, 2014. DOI 10.1109/tse.2014.2331690

Thimbleby, H. "You're right about the cure: don't do that," *Interacting with Computers*, **2**(1): 8–25, 1990. DOI 10.1016/0953-5438(90)90011-6

Thimbleby, H. "The problem isn't Excel, it's unprofessional software engineering," *British Medical Journal*, **371**:m4181, 2020. DOI 10.1136/bmj.m4181

Thimbleby, H. *Fix IT: See and solve the problems of digital healthcare*, Oxford University Press, 2021a.

Thimbleby, H. "The pivotal pandemic: Why we urgently need to fix IT," *Proceedings of the 29th Safety-Critical Systems Symposium*, M. Parsons & M. Nicholson, eds, **SCSC-161**:413–427, 2021b.

Thimbleby, H. & Cairns, P. "Interactive numerals," *Royal Society Open Science*, **4**(160903), 2017. DOI 10.1098/rsos.160903

Wallis, N. *The Great Post Office Scandal: The fight to expose a multimillion IT disaster which put innocent people in jail*, Bath Publishing, 2021.

Appendix: Touch voltage bugs

This appendix explores another safety-related bug with the Megger RCDT320 RCD tester. The appendix shows the complex relation between software and the physical reality embedded computers interact with. Many bugs are either much more debatable or much more complex to describe than the case with this example.

Electrical test instruments compliant with standards BS EN 61557, IEC 60364, etc, must check that touch voltages (and currents) will not rise to a dangerous level during a test. If touch voltages are, or are predicted to be, unsafe, test equipment must stop any test and show a warning. In other words, at least some of the technical requirements for RCD testing are clear.

Figure 3 shows simple application of an RCD tester, for instance as it might be used to check an RCD in a consumer unit. The tester will simulate a fault current, connecting line to earth. It will check the RCD trips and isolates the supply; for instance, for a fault of 150mA, a 30mA rated RCD must trip within 40ms (depending on the type of installation). But taking a test current of 150mA means that the touch voltage of the RCD (and tester for that matter[13]) would be, by Ohm's Law, 0.15A times the impedance of the earth conductor back to the electricity substation in parallel with any other places where there is a bond to earth.

As shown in Figure 3, the tester cannot measure the touch voltage directly,[14] but it can estimate it, since the voltage it can measure across its terminals L0-L1 will drop by approximately I×(Z1+Z3+Zsubstation), where I is the user-selected test fault current. The tester will estimate the touch voltage V as around 2/3 of this, using its estimate of Z3/(Z1+Z3). Note that such calculations are done by an embedded computer.

However, when testing other RCDs in an installation, such as used in outdoor sockets or extension leads, it is very convenient not to trip any consumer unit RCDs. An isolation transformer will therefore be used, as shown in Figure 4. (For clarity, Figure 4 does not show the consumer unit protections or upstream RCDs.)

An isolation transformer (such as the Seaward NTB-1) ensures no test fault current flows from line or neutral to earth (L or N to PE) on its primary side. The test fault current only flows through the secondary side of the isolator, and not through any upstream RCD that might trip and undermine test results.

With an isolator in use, it is clear that the tester now has no information at all to help it estimate Z3 (the dominant part of Z3+Z6) so it can have no idea how to estimate what the touch voltage might be.

Unfortunately, the isolator — primarily because they need to be portable and lightweight — will have a relatively high secondary impedance Z5, typically thousands of times higher than Z3. Therefore, the voltage dropped across the tester used as in Figure 4 may exceed its pre-set touch voltage limits, so the test will not operate.

In reality, the actual touch voltage is entirely dropped across Z6, but Z6 is practically zero, typically being a 13A socket/plug isolator/RCD substantial earth connection. In other words, the tester's computer program for estimating touch voltage has a bug, in that it does not correctly account for the standard testing practice of using an isolator — the whole point of using an isolator is to significantly reduce touch voltage risks.

---

[13] The touch voltage limit restricts the touch voltage of *anything*. The equivalent circuits in Figures 3 and 4 perhaps give the misleading impression that the touch voltage only applies to the RCD itself, but anything (e.g., a Class I device) on the earth fault path at that point would have similar touch voltages. The RCD tester is of course on the fault path.

[14] A few test devices do measure touch voltage directly, typically by providing a terminal the operator touches, thus providing the test equipment with an estimate of the local earth potential. Presumably such testers refuse to proceed if the terminal is not touched. (I have not had access to such a tester; and the available user manuals of the testers providing this feature do not discuss whether any such safety precautions are programmed into the tester.)

The effect of this bug is exacerbated with the RCDT320 tester as it makes none of its measurements or calculations visible to the operator, so they cannot exercise informed professional judgement.

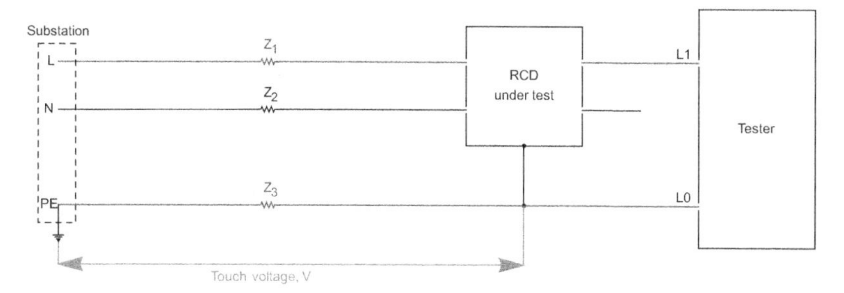

Fig. 3. An RCD tester connected to an RCD under test, for instance as might be found on an extension lead. In the circuit shown, the test fault current path is shown in blue. Zero test current flows through the neutral, N, as it is not connected, so the RCD under test registers a current difference equal to the test fault current.
(This is an equivalent circuit with Z_1–Z_3 being impedances by Thévenin's theorem. For clarity, negligible impedances are not shown.)

This bug has several consequences:

- Without an isolator, upstream RCDs may likely trip during the test, which will be an inconvenience to the user (who has to find and reset affected RCDs and consumer units), as well as to other users on the same circuit who lose power, perhaps aborting critical work.
- The user loses the protection of the isolator, including its current limit of $245/Z_5$,[15] so, ironically, if the RCD under test is faulty (including being incorrectly wired), it or any equipment connected to it could have a dangerous touch voltage.
- Unable to use an isolator, the user may be unable to test the RCD reliably, so a potentially faulty RCD may remain in service, raising future safety issues.
- If the RCD and appliance it is protecting has no exposed metalwork, the touch voltage estimate is moot anyway, especially if leakage tests have already been passed by the appliance. It is surprising that there is not a touch voltage override for this case.
- The obvious workaround is to build an isolator with a higher VA transformer to reduce Z_5 to a low-enough value that testing is possible.

---

[15] The standard mains voltage in the UK is nominally 230 V; 245 V is within range.

Home-made testing devices are unlikely to conform to safety regulations.

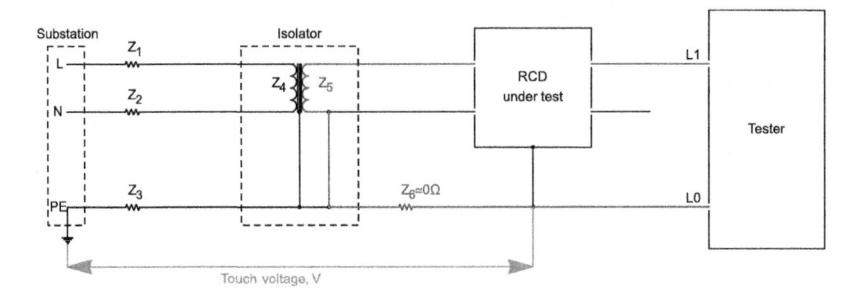

Fig. 4. Equivalent circuit of an RCD tester used with an isolating transformer. The test fault current path is highlighted in blue, creating a current differential the RCD will detect (the RCD neutral carries zero fault current, as in Figure 3). The currents through the isolator primary L and N wires are equal, so any upstream RCDs (e.g., in the consumer unit, not shown here) will not be tripped. Assuming no actual faults in the circuit, there will be zero test fault current through Z_3 so the touch voltage V is the negligible voltage drop across Z_6, which is typically the very low impedance of the RCD plug/socket earth contacts.
(As in Figure 3, for simplicity the consumer unit and other protective circuits are not shown, and negligible impedances with the exception of Z_6 are not shown.)

The fact that, when using a commercial isolator, a tester refuses to test a faulty RCD or a circuit with an incorrectly estimated high touch voltage does not make the circuit safe. Estimating touch voltage is, of course, a useful provision (in fact, it is required by BS EN 61557), but providing no possible override for when the calculation is incorrect is unsafe. However, in the use cases discussed here, an RCD tester will over-estimate the touch voltage, so it will err on the safe side for the test, but not for the requirement to test the RCD.

Could the Introduction of Assured Autonomy Change Accident Outcomes?

Dewi Daniels

    Software Safety Limited

Chris Hobbs

    BlackBerry QNX

**John McDermid,
Mike Parsons**

    Assuring Autonomy, International Programme, University of York

Bernard Twomey

    Kongsberg

Abstract *An analysis of accident reports from the air-traffic control, aviation, highways, and maritime sectors suggests areas where introduction of autonomous functionality could have materially affected the outcome. The effect of replacing or augmenting the human operator in the causal chain of the accident is considered and changes to the outcome are suggested. The functionality and assurance of the autonomous functions are considered as are the interactions with operators. Common themes are extracted from the accident studies and are used to draw conclusions about the potential benefits from introduction of autonomous functionality.*

1 Introduction

The overall aim of this paper is to see where replacing human decision making and control in a critical situation with autonomous functionality could lead to improved outcomes; it also explores the issues that arise. In many areas it is possible, indeed likely, that the introduction of autonomous functionality could materially affect the outcome of many accidents, potentially for the better. The impact of adding specific autonomous functionality to address some of the causes

---

© Dewi Daniels, Chris Hobbs, John McDermid, Mike Parsons, Bernard Twomey, 2022.
Published by the Safety-Critical Systems Club. All Rights Reserved.

and impacts of documented accidents in varied sectors is explored. The sectors considered are: air-traffic control, aviation, highways, and maritime. We believe that the outcome could have been improved in each case with the addition of credible autonomous functions. The interaction with humans is considered, as are the routine mitigations that operators utilise for day-to-day operation. Further aspects considered include:

- Human acceptability of the solution
- Robustness and resilience of the autonomy, including the flexibility and adaptability of the solution
- The day-to-day workarounds and avoiding actions that would typically be done by a human operator but are now required functions of the autonomy
- The assurance required of the proposed autonomous functions

So far as possible, we seek to be realistic and to avoid hindsight bias.

1.1 Acronyms and Definitions

AAIP	Assuring Autonomy International Programme
ABS	Anti-lock Brake System
ACC	Automatic Cruise Control / Adaptive Cruise Control
AEB	Automatic Emergency Braking, or Autonomous Emergency Braking
ALKS	Automatic Lane-Keeping System
AMLAS	Assurance of Machine Learning for use in Autonomous Systems
AS / AF / Autonomous Function	Functionality provided by a system which (i) advises a human in complex or non-obvious tasks, (ii) replaces the human-decision making in a task, or (iii) takes control of a task with no human involvement.
ATC	Air Traffic Control
FAA	Federal Aviation Administration
ICAO	International Civil Aviation Organization
IMO	International Maritime Organization
MCAS	Maneuvering Characteristics Augmentation System
MRC	Minimum Risk Condition
RIMCAS	Runway Incursion Monitoring and Collision Avoidance System
V2V	Vehicle to vehicle [communications]
V2X	Vehicle to everything [communications]

1.2 Accidents and Autonomy in the Literature

The most high-profile accident cases involving autonomy involve the Boeing 737 MAX. In these accidents an inbuilt autonomous software function (the MCAS) led to two tragic accidents by forcing the nose of the aircraft down repeatedly (SCSC, 2021). Although this might be considered an automatic function, the authors consider the distinction unimportant: the MCAS effectively took control of the aircraft from the pilots and behaved in a complex and unpredictable manner (at least to the pilots)[1] and it can be viewed as "accidental autonomy". However, whilst we are well aware of the potential and real "downsides" of autonomy, our aim here is to complement the existing literature by considering the possible "upsides".

1.3 Selection of Accident Cases

Accidents cases have been chosen on the basis of:

- Where there is significant human input, i.e. manual decisions or control of operation which led to the accident or in the failure to mitigate its consequences
- Where we think a suitable autonomous system or systems could have helped, i.e. either (i) have prevented the accident or (ii) could have mitigated the impact of the accident to a significant degree
- Where we think autonomy could credibly be introduced in the near future

1.4 Approach

The aspects considered for each autonomous function addressing the specific accident under study are given below:

1. Identification of the decisions or actions taken by humans that contributed to the accident
2. Discussion of the proposed new autonomous functionality and why it might help

---

[1] See: https://www.york.ac.uk/assuring-autonomy/news/blog/accidental-autonomy/ (accessed 23rd November 2021)

3. The data and inputs (sensors, reference information, knowledge and experience required, etc.)
4. The assurance needs of the system
5. The likely outcome of using an autonomous system in each accident scenario: in terms of causes, impact and follow-on consequences
6. The practical implications of using such an autonomous system within the sector and situation context (e.g. industry constraints, staff acceptance, training, skills, etc.)
7. The Safety-II implications of replacing human decision-making by autonomous function: the routine workaround decisions that would have to be made by the autonomous function and what skills would be lost by humans.
8. The possible new risks due to the introduction of an autonomous function, including human understanding, expectations, explanations and behaviours regarding the autonomous function
9. The limits of the proposed autonomous function functionality are explored. The issues of handover to a human operator are considered.
10. Any additional legal / liability / commercial implications for use of the autonomous function in this sector context.

2 Aviation Accident

2.1 Brief Accident Description

The facts in this section are taken from Aviation Investigation Report A14O0217 issued by the Transportation Safety Board of Canada [TSB (2016)][2].

At 09:07 EST on 11th November 2014 a 2-seater Cessna 150M aircraft, rented from Toronto Buttonville airport, Ontario, began a round trip to Peterborough (89 km), Ottawa Rockcliffe (254 km), Trois Rivières (248 km), back to Rockcliffe

---

[2] Comments are based on the author's experience as a flight instructor at the departure airport, and on his familiarity with the area where the incident occurred.

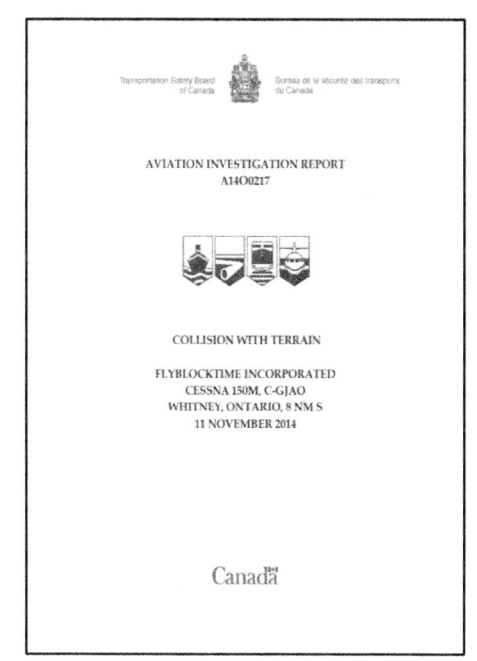

(248 km) and back to Buttonville (343 km). Without wind, this trip would take about seven flying hours.

A pilot and passenger were on board. The pilot had a private pilot licence and, with 210 hours of flight time, needed extra cross-country experience to obtain his commercial pilot licence. The first four legs of the trip were uneventful, except that at one point the aircraft entered protected airspace without permission.

On 11th November 2014, nightfall at Rockcliffe was at 17:27. The aircraft arrived at 17:26, refuelled, and departed for Buttonville at 18:03 under night Visual Flight Rules: i.e. the pilot was flying "with visual reference to the ground" rather than depending on navigation instruments. All of his previous experience of night flying had been in the well-lit area around Toronto. The agency that rented the aircraft had offered the pilot an aircraft with GPS, but he had declined this because he was not familiar with GPS.

The terrain between Ottawa and Toronto is inhospitable, sparsely settled and, at night, extremely dark. Ground-based navigation beacons (VORs) are sited in Ottawa, at Coehill, just over halfway to Toronto, and in Toronto. It is assumed that the pilot intended to fly from VOR to VOR, although this is uncertain as, contrary to regulations, no flight plan or itinerary had been filed.

The aircraft's single VOR receiver was removed and bench-checked after the accident. It initially worked correctly, but once it warmed up, it displayed an erroneous value that would cause a pilot to fly 25° too far to the right. This would take the aircraft northwards, towards the Algonquin National Park. An approaching cold front was causing the weather to deteriorate with winds forecast from the south-west — this would also push the aircraft northwards unless the pilot compensated.

The aircraft's transponder was interrogated by secondary ground radar for the first 96 minutes of the flight; contact was then lost, as is normal between Ottawa and Toronto at the altitude it was flying. While it was being tracked, the aircraft flew erratically, deviating up to 36 km north of the intended track and flying about 25° too far to the right: see figure 1.

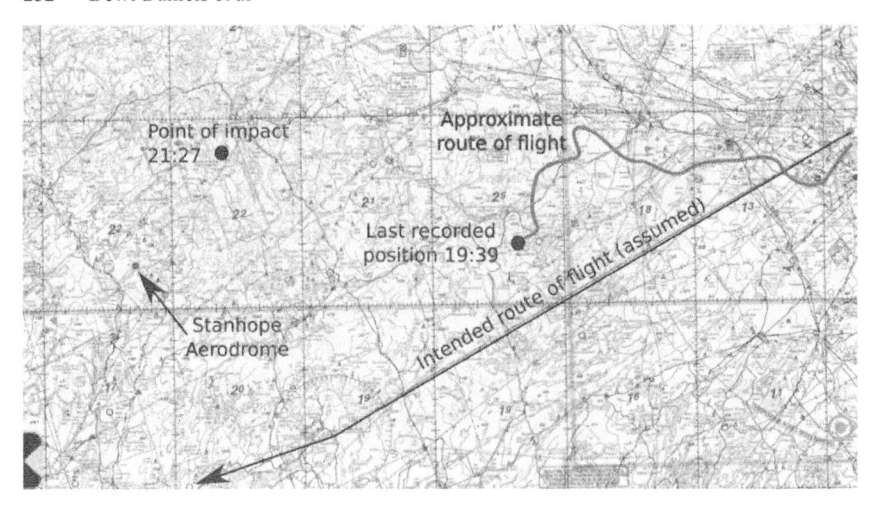

Fig. 1. Aircraft Track Using Transponder Data

At 20:25 the pilot made a Mayday call. The aircraft was too low for this to be received by a ground station, but it was relayed to Air Traffic Control (ATC) by an overflying commercial aircraft. Through various overflying aircraft, ATC tried to locate the aircraft by asking the pilot to take bearings on the Coehill VOR. The pilot reported that he was on the Coehill 170° radial (i.e., south of it), when in fact he was significantly north of it.

At 20:41 ATC instructed the pilot to climb so that his transponder could be interrogated. However, a cloud layer prevented this: the pilot was not certified to fly in cloud, and his training had stressed that the disorientation associated with entering cloud was normally fatal for an untrained pilot.

At 20:51 the pilot told ATC that his fuel tanks were "almost empty" although he still had an hour's worth of fuel. Estimating the aircraft's position, ATC gave the pilot a heading to fly to reach the nearest lit airport.

At 21:27 the pilot reported that the aircraft was descending, and a crash followed about 75 km north of the intended track. The engine was turning at impact and there was fuel on board for another 30 minutes of flight.

A search and rescue helicopter reached the site crash at 23:12. Low cloud and blowing snow prevented it from lowering crew members until 03:00. They then found the two occupants of the aircraft had not survived and that neither had been wearing the available shoulder harnesses.

2.2 Identification of Key Decision / Action Points

The key decision point here was the pilot's decision to take off from Rockcliffe, after dark, while tired and presumably hungry (the only food available at

Rockcliffe was from a vending machine). That decision was probably affected by the desire to get home and the confidence raised by the successful completion of the first four legs.

After the final flight started, the key decision point was when the pilot determined that he needed help and called Mayday.

In summary, an inexperienced pilot reached a situation where he was lost, exhausted, probably hungry, flying in conditions that he had never met before, with an incorrect knowledge of his remaining fuel, relying on an inaccurate and antiquated navigation device. He did not use the aviation application on his mobile telephone which could have provided a moving map display to his destination using GPS. He did not follow the magnetic compass south which would have brought him into a more populated and better lit area.

This incident poses two questions: could an autonomous system have prevented the pilot from reaching this situation, and could it have prevented the accident once the situation had been reached?

2.3 Proposed New Autonomous Function(s)

The first opportunity to avoid the situation was before leaving Rockcliffe. An autonomous device knowing the history leading up to the departure would have been able to advise strongly against taking off, but it is unlikely that the pilot would have been willing either to enter sufficient information to make such a device useful, or to take its advice: after completing four flights successfully, he was operating within his envelope of comfort, although outside his envelope of ability.

However, there was one agency with a pecuniary, as well as moral, incentive to have the flight completed successfully: the company that owned the aircraft. Had such a device been available, the dispatcher could have programmed it before the departure from Buttonville with the intended route of flight, the previous experience of the pilot and the type of aircraft. Using GPS, the device could determine its position, and using satellite or cellular connections it could monitor weather forecasts. It could have warned the pilot, and more importantly could have sent a message to the rental agency after the aircraft landed at Rockcliffe, recommending that the flight back to Buttonville be postponed. This could have allowed a flight instructor to contact the pilot and discuss the situation.

The second opportunity to prevent the situation was during the first 96 minutes of the flight during which ATC's secondary radar was tracking the aircraft. Because the pilot had not filed a flight plan, the identity of the aircraft was unknown, but its position and (unverified) altitude was known. The recorded speed would indicate that it was some form of light aircraft. Such an erratic track during the day would not have been cause for alarm, but at night, with deteriorating weather, the track was unusual. ATC systems are programmed to alert an operator to certain conditions (e.g. two aircraft on tracks which will cause loss of separation), but these are primarily computational rather than inferential systems.

An ATC-based inferential system could have detected the unusual behaviour of the aircraft, given the radar track (see figure 1), the lack of a flight plan, the time of the year and the clock time, and ATC could then have tried to make contact with the pilot on one of the radio frequencies that a pilot is advised to monitor while *en route*. The pilot could then have been issued with a specific transponder code and, while remaining in radar contact, could have been guided to a nearby airport. It is psychologically likely that, at this point, the pilot would have accepted assistance.

Between the Mayday call at 20:25 and the crash at 21:27 the pilot clearly knew that he needed help but did not think to use the application available on his mobile telephone or simply head southwards towards more populated areas using the magnetic compass. However, at this time, he probably would have been willing to activate a device with a "Help Me!" button.

2.4 Information Required

The basic information needed by such a device carried in the aircraft would be the circumstances of the flight: the type of aircraft, the intended route, the available navigation equipment, the experience of the pilot, etc. Given that the pilot did not even file flight plans, it is unlikely that he would have taken the trouble to enter this information, but it was in the interests of the dispatcher to do so. Once airborne, additional information could be accessed: location of the aircraft, time and date of departure, weather forecasts, etc.

An application looking for unusual tracks on ATC's secondary radar would require only the radar track and the time and date.

2.5 Assurance

Garmin has announced its Autoland feature[3] which will take control of a suitably-equipped aircraft if the pilot is incapacitated, find the closest suitable runway and land the aircraft autonomously.

In contrast, the device proposed here acts as an advisor and, by instant messages, incorporates experts on the ground in the advice it gives to the pilot. Its intention is to avoid an accident, not to mitigate the effect of the accident. In this

---

[3] https://www.planeandpilotmag.com/article/piper-m600-sls-the-first-production-plane-that-lands-itself/

role, a high degree of assurance is required only in its logic for detecting an anomalous condition, where a false negative could be dangerous.

One advantage the device has is that the air is a much emptier environment than a road and the simulated conditions to demonstrate the necessary level of assurance would be much simpler than for a road vehicle.

2.6 The likely outcome

Given the device in the aircraft, a warning could have been issued while the aircraft was on the ground at Rockcliffe, recommending that, given the forecast weather, the experience of the pilot and the limited navigation equipment available, the trip be postponed to the following day. It is likely that the pilot would have ignored this: his envelope of comfort was outside his envelope of ability.

However, once the pilot was in the air, the device could have warned the pilot that he was deviating significantly from the desired track and recommend he contact ATC (giving the radio frequency) or propose a track to fly. It is highly likely that, before radar contact was lost at 19:39, the pilot would have taken this advice - he was already unsure of his position.

Particularly if the device had been pre-programmed by the dispatcher before leaving Buttonville, it could have provided ATC and the rental agency with instant messages (through the Iridium satellites) giving the precise location of the aircraft. Given the date, time and detailed terrain data (available in GPS databases), such a device could have known that the neighbouring lakes were not yet frozen (and therefore were not feasible landing sites), that there were few roads in the vicinity, and that the nearest airport with lights was Stanhope. If the device further knew the time of takeoff (deduced from the aircraft's speed first exceeding 50 km/hr) and could assume full tanks at that time, it would be able to estimate whether Stanhope could be reached and tell the pilot how to activate the lights there (this requires that the radio be tuned to a particular frequency and the push-to-talk button be pressed a defined number of times). If Stanhope could not be reached, the device could at least identify the best amongst the other options for landing areas to the pilot.

It is likely that such a device could have saved the pilot and his passenger.

2.7 The practical implications

The device proposed introduces no new technology: several devices exist that send and receive emails and text messages via the Iridium satellites; every mobile telephone has a GPS receiver, clock and calendar; even the simplest handheld avionics GPS receivers contain a terrain database, etc.

What needs to be added is an inference engine based on expert inputs from pilots experienced in flying in the geographical area where the crash occurred.

2.8 The Safety-II implications

As the device proposed here is advisory, it carries no negative Safety-II implications[4].

2.9 The possible new risks

The primary risk associated with adding autonomy to a currently manual system, is that it de-skills the human operators. This has been seen in the fly-by-wire systems in modern airliners, but in this case, it is unlikely that any individual pilot will become reliant on the proposed systems.

2.10 The limits of the proposed functionality

The proposed systems are advisory, not controlling. Adding control equipment, such as an autopilot, to an antiquated (1976) light aircraft would be prohibitive both in cost of installation and in obtaining regulatory approval.

2.11 Any Additional Considerations

This accident is particularly difficult to understand because the pilot had everything needed to avoid the situation: a GPS application on his mobile telephone and a magnetic compass in the aircraft. The problem was not that the accident was unavoidable, but that the pilot needed prompting to use those assets. An autonomous device, backed by experienced pilots at the rental agency, would need only to act as a nudge.

There is nothing required in either the ATC software or the device within the aircraft that is not currently available. However, the market for the proposed device in Canada would be very small and so it would need to be suitable for use anywhere in the world.

---

[4] However we know how trust changes behaviour. So just because something is advisory does not mean that it won't change risk as undue reliance can occur. So this is a possible implication.

3 Air Traffic Control Incident

3.1 Brief Accident Description

On 20 July 2020, a Boeing 787-10 operated by United Airlines was landing at Paris-Charles de Gaulle (CDG) airport (BEA 2021).

At the time, Paris-Charles de Gaulle airport was using two parallel runways; 09L was being used for landings, 09R was being used for take-offs.

The air traffic controller mistakenly cleared the United flight to land on runway 09R rather than 09L. At 05:16:48 UTC, the controller told the United pilot, "United 57 bonjour, number one for 09 Right cleared to land 09 Right, wind 010 degrees 9 knots gusting 21 knots".

The United pilot, who had been expecting to land on runway 09L replied, "Understand cleared to land 09 Right, sidestep for 9 Right United 57". The United pilot changed course to land on 09R.

At 05:17:23 UTC, the controller then cleared an EasyJet Airbus A320 to line up and wait on runway 09R. As the EasyJet flight entered runway 09R, the EasyJet pilot spotted the United aircraft turning to land on runway 09R. The EasyJet pilot braked to a halt, but the aircraft had already entered the runway.

At 05:18:10 UTC, the EasyJet pilot radioed "Tower, there is a traffic landing 09R" immediately followed by "Go around 09R, go around".

- The United pilot replied they were going around and initiated a go-around.
- The controller confirmed the go-around. At the same time, the controller's Runway Incursion Monitoring and Collision Avoidance System (RIMCAS) issued a belated warning.
- The United flight reached a minimum height of 80 ft and passed over the EasyJet flight at a height of 300 ft.

- On the second attempt, the United flight landed successfully on runway 09L.

This was potentially a very serious accident. The two aircraft had a combined total of 228 passengers and crew on board.

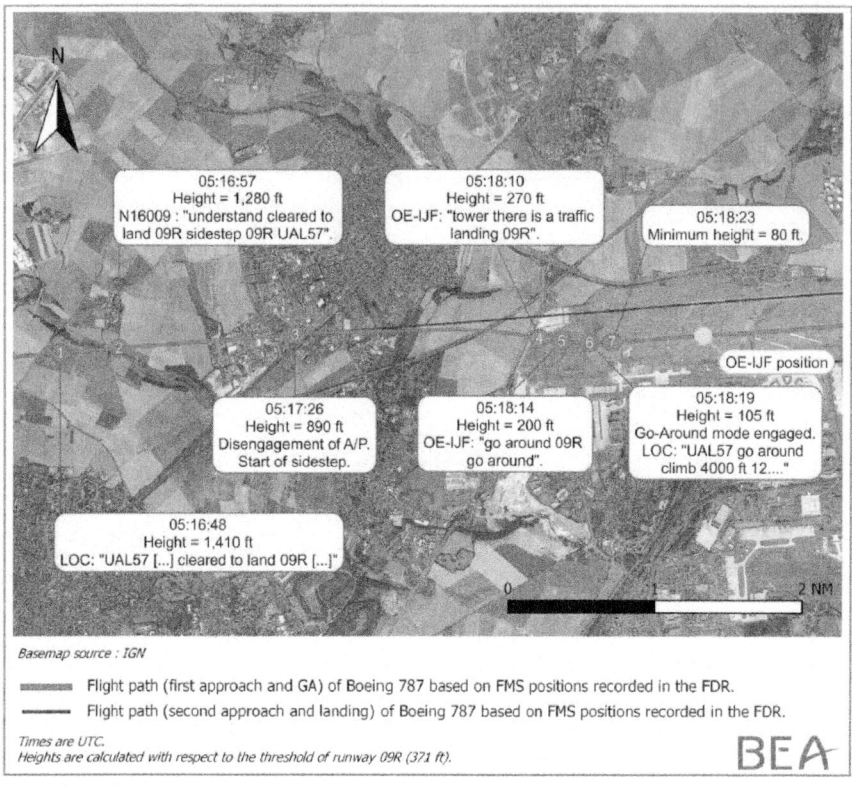

Source: BEA

Fig. 2. Path of Boeing 787 based on FDR (reproduced from the BEA investigation report)

Runway incursions are very dangerous. The deadliest accident in aviation history remains the collision of two Boeing 747 airliners on the runway at Tenerife in 1977 (Wikipedia, 1977).

3.2 Identification of Key Decision / Action Points

The main problem was the air traffic controller confusing runway 09R with 09L. Although an automated system (RIMCAS) issued a warning, the warning came a little late, given the United flight had already initiated a go around by that time, during which it reached a minimum height of 80 ft.

Fig. 3. United Airlines, N16009, Boeing 787-10 Dreamliner, Anna Zvereva, https://creativecommons.org/licenses/by-sa/2.0/

3.3 Proposed New Autonomous Function(s)

An autonomous system could have been monitoring ATC communication. When the air traffic controller cleared the EasyJet flight to enter runway 09R, the autonomous system could have alerted the controller that they had already cleared the United flight to land on the same runway. The autonomous system could be integrated with the existing automated system (RIMCAS) to use the same warning system.

240 Dewi Daniels et al

Fig. 4. EasyJet Europe, OE-IJF, Airbus A320-214, Anna Zvereva, https://creativecommons.org/licenses/by-sa/2.0/

This would have resulted in the warning being issued at 05:17:23 UTC rather than at 05:18:19 UTC, i.e., 56 seconds earlier. The United flight would still have been at about 930 ft, before it started to turn towards runway 09R.

The following table shows the time when the autonomous function could have issued a warning, the time when the EasyJet pilot issued a warning and the time that RIMCAS issued a warning.

Table 1. Timeline

Time	Event	Height	Distance to threshold
05:16:48	Controller: "United 57 bonjour, number one for 09 Right cleared to land 09 Right, wind 010 degrees 9 knots gusting 21 knots"	1,410 ft	
05:16:57	United: "Understand cleared to land 09 Right, sidestep for 9 Right United 57"	1,280 ft	
05:17:23	Controller clears EasyJet flight to line up and hold on runway 09R at holding point D5		
05:17:23	Autonomous function could have issued a warning at this point		
05:17:26	United flight turns towards 09R	890 ft	2.2 NM
05:17:50	EasyJet flight starts taxiing to line up from D5	595 ft	1.3 NM
05:18:10	EasyJet: "Tower, there is a traffic landing 09R"	270 ft	
05:18:14	EasyJet: "Go around 09R, go around"	200 ft	0.42 NM
05:18:19	RIMCAS issued a warning	105 ft	0.17 NM
05:18:19	"UAL 57 go around climb 4000 ft 12…"	105 ft	0.17 NM

3.4 Information Required

The autonomous system would need to incorporate natural speech recognition. This is made easier by the ICAO-standard terminology used in ATC communication. Also, voice recognition is now a widely available technology, e.g. through systems such as Alexia and Siri, so there is technical basis on which to build (even if it is unassured).

The autonomous system would also need knowledge of the airport layout (runways and taxiways) and of the position of the aircraft.

3.5 Assurance

The autonomous system proposed here is an advisory system. It would bring the conflict to the attention of the ATC controller, who would then take appropriate action. The interface to the ATC controller would be very similar to the existing warning system, but the autonomous function would be able to issue warnings earlier. One challenging issue would be the assurance of the voice recognition system. Even though it is advisory, frequent false alarms would lead to the system being ignored. Whilst we are not aware of work on assurance of such speech recognition systems the AAIP is sponsoring work looking at assurance of voice analysis software used for critical applications[5].

3.6 The likely outcome

In this case, the existing automated system (RIMCAS) issued a warning, but only at a late stage, and only after the EasyJet pilot had already spotted the conflict and warned the United pilot.

The autonomous function we propose could have issued a warning 56 seconds earlier, when the United flight would still have been at about 930 ft, and before it started to turn towards runway 09R.

---

[5] See: https://www.york.ac.uk/assuring-autonomy/demonstrators/ai-ambulance-response/ (accessed 23rd November 2021)

Fig. 5. Charles de Gaulle International Airport. from above (boosted colors), David Monniaux, https://creativecommons.org/licenses/by-sa/3.0/deed.en

3.7 The practical implications

One potential issue is the use of non-ICAO-standard terminology, which is particularly prevalent in North America, and which would make natural speech recognition more difficult. The French investigation report criticised the United pilot for using the word "understand", rather than the ICAO-standard "confirm". This seems a little unfair given that "confirm" appears in the ICAO Manual of Radiotelephony but does not appear in the French phraseology for general air traffic training manual (or the FAA pilot/controller glossary). It is, however, the case that the consistent use of ICAO-standard terminology would improve comprehension by pilots whose first language is not English as well as by natural language processing software.

At European airports that have parallel runways, it is conventional to use one runway for landing and the other runway for take-off. In the USA, it is common to use both runways for both take-offs and landings. This was also a contributing factor to this incident. The previous aircraft had requested to land on 09R (which is longer than 09L) due to a technical problem. This had perhaps placed 09R in the controller's mind, resulting in their clearing the United flight to also land on 09R when they meant 09L. The United pilot, who was on approach to 09L, was not as surprised to be asked to change runway as a European pilot would have been, since this is apparently common practice in the USA, even at a late stage in the approach. Nevertheless, the United pilot did query the landing clearance, albeit using non-ICAO-standard terminology.

Another issue that is not relevant to this incident but that would affect the design of the autonomous function is that it is common in North America to clear an aircraft for landing while the runway is still occupied by a previous aircraft. For example, a controller might say "United 57 cleared to land runway 09L, number two behind the Gulfstream". In Europe, the controller will more likely delay the landing clearance in such a situation, so that sometimes the landing clearance is not issued until the landing aircraft is on final approach. Both practices are likely to hamper the autonomous function's ability to issue an early warning.

4 Highways Accident

4.1 Brief Accident Description

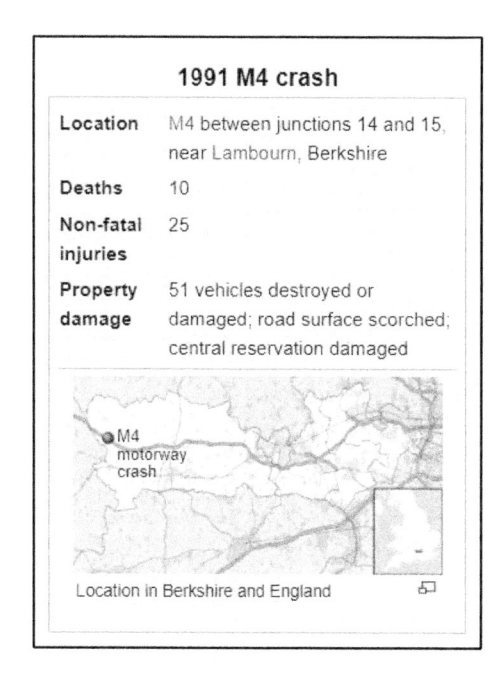

1991 M4 crash	
Location	M4 between junctions 14 and 15, near Lambourn, Berkshire
Deaths	10
Non-fatal injuries	25
Property damage	51 vehicles destroyed or damaged; road surface scorched; central reservation damaged

Location in Berkshire and England

On 13 March 1991, a multiple-vehicle collision occurred during foggy conditions on the eastbound carriageway of the M4 motorway near Hungerford, Berkshire, between the Membury service station and junction 14 (Wiki, 2021).

Ten people were killed in the pile-up, which involved 51 vehicles, making it one of the deadliest crashes in the history of Britain's motorways. A van driver claimed something flew up in front of him, startling him and causing him to skid into the central reservation, but it was later reported that he fell asleep at the wheel. A car travelling behind the van changed lanes to avoid contact but other vehicles behind, which were travelling at speeds averaging 70 miles per hour, failed to avoid the crashed van and skidded into the other lanes of the carriageway. Others took evasive action by driving onto the hard shoulder and up the sides of the cutting. An articulated lorry then jack-knifed across all three lanes of the eastbound carriageway.

One driver freed himself from his car and ran back down the central reservation to warn approaching motorists but was ignored. In a period of 19 seconds, 51 vehicles became involved in a pile-up. Fuel exploded along with the highly combustible material being carried in one of the vans (possibly deodorant) and the resultant series of explosions closed the carriageway for four days as the charred wrecks were removed and the road surface replaced. A recorded video of the accident can be found at (CorwinBrett, 2016) and a narrated account on (YouTube, 2019)

Fig. 6. Aftermath of the Accident
(http://img.photobucket.com/albums/v645/zloty/fog1991013.jpg)

There were three minor collisions caused by distracted drivers on the opposite carriageway of the motorway.

The crash led to warning signals being introduced on British motorways to warn drivers of fog.

It should be emphasized that this accident was an extreme case, but smaller accidents happen all the time on UK roads[6].

---

[6] It was noted at the time that there were more people killed and injured in this motorway crash than in the British Army from enemy fire in the Gulf War.

4.2 Identification of Key Decision / Action Points

Fundamentally, a driver swerving caused multiple delayed reactions and over-reactions by following drivers, leading to impacts and subsequently multiple deaths. The weather conditions were poor, vehicles were travelling too fast for the conditions, and vehicles were too close to the vehicles in front.

The steps in this accident can be seen in the videos and animations. In summary:

1. There was patchy fog on the motorway, but no warnings were given
2. A van driver fell asleep at the wheel or had a lapse of concentration
3. The van swerved towards the central reservation
4. Too many drivers were driving too close and too fast
5. Drivers did not react quickly enough or in an appropriate manner
6. Evasive actions, including rapid swerving across lanes, caused a chain reaction of poor reactions in other drivers
7. Multiple impacts of other vehicles
8. A lorry crashed into the back of the pile-up
9. The lorry was carrying inflammable loads which then ignited

4.3 Proposed New Autonomous Function(s)

There are many automatic systems available today in modern cars which could have prevented this accident, and there are some systems which are possible though not yet in place[7].

- There are driver sleep/attention prompter systems which can monitor the drivers state (Driver Monitoring Systems, Driver Monitoring Camera). These can detect eye movement or other driver attributes and can sounds alarms, shake the steering wheel, vibrate the driver seat, etc. These systems are currently available as options on high-end vehicles. Of course, there is no guarantee that these systems will wake or refocus the driver, or indeed be of any use if the driver has, say, an undiagnosed medical condition.
- Communication between vehicles systems – V2V (vehicle to vehicle) or V2X (vehicle to everything) systems exist as concepts and ideas. These could have communicated the crash to approaching vehicles, warning them

---

[7] Note brand differentiation introduces a range of terms and the names for all these functions are not yet fully established.

to slow down. Standards are being worked on for communications, and these are seen as a prerequisite as otherwise communications effectiveness will be limited. It is clear that a critical mass of vehicles need to have systems fitted for this to be of any use.

- Auto-detect fog systems are possible but not yet commercially available. Note that radar can see through fog, but optical cameras cannot. It is recognized that these sorts of systems could create more problems in mixed mode traffic where vehicles without the technology do not slow down[8]. There could be better and more frequent gantry signage systems to indicate fog. (Note that fog is only one of many things which may impact visibility: others include snow, low cloud, smoke, dust clouds and even sandstorms.)
- Advanced signage on a motorway could be implemented with better displays, notifying drivers approaching an accident or a hazardous stretch of road. Gantry signage could be re-displayed within vehicle (e.g. on head-up displays) and the driver specifically alerted. The information on gantries could be communicated directly to vehicles (V2X) and even into speed limiters (see below).
- Adaptative Cruise Control (many different names) has been available for some years and can maintain safe distancing between vehicles in certain situations. This can be using either time / distance separation or physical distance separation. This can utilise lidar, radar and/or camera-based systems.
- Automatic braking systems, typically Automatic Emergency Braking, or Autonomous Emergency Braking (AEB), are currently available and can apply the brakes to avoid collisions if the driver takes no preventative action. These can also pre-tension seat belts, warn the driver, etc. as part of pre-crash activities. These are already in place in many vehicles. Note that it may be possible to override the system using pedals in some cases.
- Weather/fog awareness system using vehicle-based sensors. It would be possible for vehicles themselves to report adverse weather conditions to a central repository and then this information could be distributed to all. V2V and V2X would be essential for communications. Also using multiple vehicle sensors is likely to give a more accurate and detailed result in comparison

---

[8] Or indeed those with the technology continue at speed (as they are not hampered) whereas those under manual control may reduce speed sharply.

with road infrastructure sensors. Collected V2X data could also be very useful for road maintenance[9].
- Speed limiters could be automatically imposed related to weather or traffic conditions. Automatic Cruise Control (ACC) can set vehicle speed today, but it can be overridden. Again, mixed traffic is a problem, with connected vehicles responding to speed limits whereas human-driven vehicles would likely not. If all vehicles were autonomous and connected then it could work[10].
- There is a question as to whether Automatic Lane-Keeping System (ALKS) would have helped: in the accident, some drivers swerved abruptly to avoid other vehicles, and this seemed to have benefit (at least to that driver). We have to assume that with ALKS the driver would take some time to gain situational awareness and that could have made things worse.
- Some other systems which may have helped include Head-Up Displays (HUD), radar displays and Augmented Reality.

Further information on proposed vehicle safety systems may be found at (RAC, 2017). It is noted that AI may help to predict accident black spots, e.g. deep learning helps predict traffic crashes before they happen e.g. (MIT, 2021).

Fully autonomous vehicles (no driver) would incorporate some of these technologies as well as other higher-level functions. It is assumed such vehicles and would be fully internet-connected at all times, likely supported with a remote monitoring and control centre. In this case they would avoid the accident area as soon as notified; they would also keep a safe distance and be aware of weather conditions.

Whether they would cope with the mixed traffic scenario (some manual, some autonomous) is unclear; how they would cope with other drivers breaking the rules of the road, e.g. to swerve across lanes, and what specific avoiding actions they could take when, e.g. a vehicle close by bursts into flames, is unknown.

---

[9] The "vehicle becomes the sensor". Data collection could be from many vehicle systems including automatic headlights, braking lights, automatic windscreen wiper operation along with speed of the vehicle. This could alert the control room operators to a number of conditions which could potentially be more effective in managing safety. Using the vehicle as a sensor not only provide a more connected transport system but could also pick up the 'odd roadworks' you find when being diverted.

[10] If the cars are interconnected and provide information to other road users of worsening environmental conditions, or reductions in speed then that could be of a significant benefit to other road users.

4.4 Information Required

For all the above systems, real-time data is required, mostly from vehicle sensors but also from infrastructure systems and internet providers. This data becomes safety-critical when used to slow, stop the vehicle, or take control to avoid accidents[11]. Of course, some of these sensors may be faulty, missing, or, for vehicle sensors, disabled by the driver. The data must be processed in real-time and decisions made in fractions of a second. Other factors such as the proximity of pedestrians, motorcycles and vehicles immediately behind need to be incorporated into the overall picture.

4.5 Assurance

Assurance is required for all systems which perform a safety function: the assumption that the driver is in control doesn't apply to advanced automatic and autonomous functions, and the idea that a driver can take over control of the vehicle in all situations within the time available to avoid accidents is currently under study[12].

Software assurance will be required using e.g. ISO 26262 (Road vehicles — Functional safety) and ISO/PAS 21448 (Road vehicles — Safety of the intended functionality). New regulations currently under development for autonomous functions will have to be adhered to. Security will have to be addressed by ISO/TR 4804 (Road vehicles — Safety and cybersecurity for automated driving systems) or similar. There are also regulations for specific functions, e.g. ALKS (UNECE 157). New techniques such as Assurance of Machine Learning for use in Autonomous Systems (AMLAS) (AMLAS, 2021) will be needed for the machine learning components of autonomous functions.

4.6 The likely outcome

It is clear that many of these systems could and probably would have helped – an emergency braking system looking ahead with radar, etc. – and likely avoided the accident for at least some of the vehicles. Better warnings and guidance transmitted between the vehicles could have provided advance information.

---

[11] The guidance from the SCSC on Data Safety may be useful, (DSIWG, 2021).
[12] For instance (TRL, 2020) "Safe performance of other activities in conditionally automated vehicles: Automated Lane Keeping System" states *"...it appears sensible to conclude that tasks which require significant visual attention will impact the ability of drivers to safely resume control of the vehicle"*

Of course, the fundamental cause here was that the traffic was too close together and going too fast to reasonably avoid the accident, hence better information about the conditions ahead on gantries or head-up displays could have warned drivers, and systems such as adaptative cruise control could have increased separation.

4.7 The practical implications

To have any effect on accidents such as these, the deployment and adoption of these systems within vehicles must be high (i.e. installed on the majority of vehicles on the road). Moreover, drivers must trust these systems and not disable them. An example is the introduction of running lights in the UK in 2011[13] which was followed by a series of questions on the internet from vehicle owners asking how to switch them off.

4.8 The Safety-II implications

There are some serious questions about introducing such automatic and autonomous functions into vehicles as there is a real possibility of making drivers lazy[14], losing skills of the road, or worse, being incapable of taking control in an emergency situation. There are also big and unsolved problems with mixed mode traffic on the same roads (some manual, some with partial automated functions, some fully autonomous). How will these vehicles interact safely?

There is no doubt that drivers will start to lose skills as soon as these systems become commonplace. Do drivers need more regular training and testing to keep their skills current? At the very least, the driving test may need to include a section on emerging technologies and how to use them.

There is a concern that drivers may also start to "game" autonomous functions turning them on and off to gain advantage on the road in specific situations.

---

[13] www.gov.uk/government/publications/daytime-running-lights/daytime-running-lights

[14] A simple example is automatic switch-on of headlights, featured in many modern cars. If the driver then uses an older car they may forget to switch on the lights.

4.9 The possible new risks

There are significant new risks with these vehicle systems, most of which have been covered above already. There is also a risk that the new systems do not work well together, do not communicate properly with other road users or a central control, switch modes too often, are poorly understood by the driver (and other road users) and cause confusion. Further, road traffic should be viewed as a 'system of systems' (SoS) and hazard and safety analysis undertaken accordingly; whilst there is research on SoS analysis, this is not a well-established area of functional safety.

There is also individual risk with each new system: the level of sophistication prevents simple analysis and any form of exhaustive test, so much verification will be left to analysis and simulation. Machine learning components do not yet have an accepted assurance regime, although some are now proposed (AMLAS, 2021).

4.10 The limits of the proposed functionality

Currently the available automatic systems can hand control back to the driver although it could be argued that ALKS and AEB do, in fact, replace the driver to an extent. There needs to be a clear position on what is reasonable to be handed back to a driver, given limited time to take back control of the vehicle, and the systems should also be designed to 'remain safe' if the driver does not regain control, e.g. by bringing the vehicle to a minimum risk condition (MRC).

4.11 Any additional Considerations

There are legal and liability implications for the use of the autonomous functions in the automotive sector where the driver can no longer be said to be in complete control. The Law Commission is currently looking into this, with this remit, reporting back at the end of 2021 (LawCommission 2021):

> *"The Centre for Connected and Autonomous Vehicles (CCAV) has asked the Law Commission of England and Wales and the Scottish Law Commission to undertake a far-reaching review of the legal framework for automated vehicles, and their use as part of public transport networks and on-demand passenger services.*

Could the Introduction of Assured Autonomy Change Accident Outcomes? 251

By automated vehicles we refer to vehicles that are capable of driving themselves without being controlled or monitored by an individual for at least part of a journey. We also consider issues arising at the boundary between self-driving vehicles and widely used driver assistance technologies such as cruise control."

Many of these functions are already available on high-end modern vehicles. Some functions are further away; infrastructure will likely take longer to develop as this is not subject to the commercial pressures present on vehicle development.

5 Maritime Accident

5.1 Brief Accident Description

The factual information in this section is taken from the Ministry of Infrastructures and Transports - Marine Casualties Investigative Body - Cruise Ship-COSTA CONCORDIA - Marine casualty on January 13, 2012.

"On 13 January 2012, whilst the Costa Concordia was in navigation in the Mediterranean Sea (Tyrrhenian sea, Italian coastline) with 4229 persons on board (3206 passengers and 1023 crewmembers), in favourable meteo-marine conditions, at 21 45 07 LT (local time) the ship suddenly collided with the "Scole Rocks" at the Giglio Island. The ship had just left the port of Civitavecchia and was directed to Savona (Italy).

The ship was sailing too close to the coastline, in a poorly lit shore area, under the Master's command who had planned to pass at an unsafe distance at night time and at high speed (15.5 kts).

The vessel immediately lost propulsion and was consequently affected by a black-out. The Emergency Generator Power switched on as expected but was not able to supply the services required to handle the emergency.

The rudder was blocked and was completely over to starboard, at the Giglio Island at around 23.00.

From the analysis carried out under the direct coordination of the Master, the seriousness of the scenario was reported after 16 minutes. After about 40 minutes (22:27) the water reached the bulkhead deck in the aft area. The assessment of the damage was continued by the crew, realizing, at the end, that watertight compartments (WTC) nos. 4, 5, 6, 7 and 8 were involved.

A combination of factors caused the immediate and irreversible flooding of the ship beyond any manageable level. The scenario of two contiguous compartments (WTC 5 and 6) being flooded and the period of time after contact for the flooding to occur already represented a limited condition as far as buoyancy, trim and stability are concerned to allow safe and orderly evacuation of the ship.

The ship's stability was further hampered by the simultaneous flooding of another three contiguous compartments, namely WTC's 4,7 and 8. The flooding of these additional compartments increased the ships draught so that Deck) (Bulkhead deck) started to be submerged. The effects of the free surface created in these compartments occurred in about 40 minutes, which resulted in the first significant heeling to starboard.

At 22:54:10 the abandon ship was ordered, but it was not preceded by an effective general emergency alarm (several passengers – in fact - testified that they did not catch those signal-voice announcements). The first lifeboats were lowered at 22:55 and at 23:10 they moved to the shore with the first passengers on board.

Crewmembers, Master included, abandoned the bridge at about 23:20 (only one officer remained on the bridge to coordinate the abandon ship).

At about 24:00 the heeling of the vessel seriously increased, reaching a value of 40°. During the rescue operations it reached 80°.

At 00:34 the Master communicated to the SAR Authorities that he was on board a lifeboat with other officers.

All the saved passengers and crew members reached Giglio Island (the ship had grounded just a few meters from the port of Giglio). First rescue operations were completed at 06:17, saving 4194 people.

The rescue operations continued, and on 22nd March the last victim was found. The number of victims were 32, 26 passengers and 4 crewmembers died in this event".

Fig. 6. Costa Concordia Accident

(Rvongher / Wikimedia Commons, CC BY-SA 3.0 <https://creativecommons.org/licenses/by-sa/3.0>, via Wikimedia Common)

5.2 Identification of Key Decision / Action Points

It has been concluded from the safety investigation that human error was a significant contributory factor to the accident. Costa Cruises CEO, Pier Luigi Foschi, explained that the company's ships have pre-programmed routes and "alarms for both visual and sound, if the ship deviates for any reason from the stated route as stored in the computer, and as controlled by the GPS", but these alarms could be "manually" overridden".

The report also states that the "appropriate navigational charts were not on board the vessel", therefore the navigational hazards were unknown for the deviated route.

> *"The area where the accident occurred is covered by charts of the Hydrographic Institute 6 - 1:100,000 - and 119 - scale 1:20.000 - (Isola del Giglio). The nautical chart appropriate for the planning and monitoring of navigation in the proximity of the island is therefore 119.*
> *The ship, as it turned out by the inventory of charts (Annex 26), relative to the area of the accident, was not equipped with the 119 chart for navigation near the island of Giglio.* ***This is acceptable*** *because the navigation near the island of Giglio was not scheduled routes normally used by in the ship".*

In this case, the lack of navigational charts for the route, the fact that alarms and warnings can be overridden and final decision to proceed is that of the master, indicates that the current arrangements would not prevent another accident with a similar signature from occurring.

5.3 Proposed New Autonomous Function(s)

The use of an AI system that alerts the operators of a potential hazard whilst making the decision open for human review and scrutiny, could be introduced as an advisory function. This function would require the operators to understand the information being presented and accept or reject the advice. This is no different to the current systems already provided on-board as the final decision to proceed is that of a human.

The report made a number of recommendations, despite the human element being the root cause in the Costa Concordia casualty.

From the report:

After this investigation, there is the opportunity to deliver in the hands of the International Maritime Community some suggestions regarding as the naval gigantism, represented in this case by the Very Large Cruise Ships, to face this actually and rising wonder through to the following items should be focused systematically also in the future:

Recommendations:

- mitigate the human contribution factor with education, training and technology;

We know that technology is outstripping the educational qualifications of the crew. A ship can now be considered as a complex, highly coupled set of elements which should interact according to design. These elements can include hardware, software and the human element.

Questions:

1. Do we try to educate the crew to work within a highly coupled complex software-based environment, or provide a system that assists them in understanding the capabilities and limitations of the system?
2. Could an autonomous function reduce the requirement to have suitably qualified and experienced personnel interacting with a complex highly coupled software-based environment?

3. What educational qualifications would be required for seafarers of the future? If the educational standards increase, where will we get these people from?[15]

- operate day by day directly to support the shipping industry (shipbuilding), investing in the innovation technology;

Increasing the use of technology without understanding the risks is not going to work. The industry must follow a systematic through life process for risk identification and mitigation at the infrastructure level. This will be challenging, as not all stakeholders are at the same level of understanding on how to achieve this requirement. To do this the industry will need to look at all aspects of their business, and develop an assurance framework that is not solely based on prescriptive requirements.

- stress all the maritime field cluster to make the maximum contribution for the related study and consequent technical research.

It is hoped that the maritime industry will learn from other sectors, and not try to 'reinvent the wheel'. If autonomy is introduced, an assurance process that addresses non-deterministic systems needs to be developed which includes taking into consideration the role of the regulatory authorities and legal framework under which the maritime industry operates.

Q. Could an AI system have alerted the navigational crew that they did not have the appropriate charts for the voyage?
A. The answer is yes, but as stated by Mr Foschi, deviation alarms can be manually overridden and information provided by an AI system could be ignored.

Q. Could the AI advise the crew that they were deviating from an approved route?
A. Yes, but the vessel already has a method of alerting the navigating officers if they are deviating off course.

Q. Could an AI system prevent the vessel from following a route that could result in an; accident taking place?
A. Yes, but this may be quite controversial, as the AI could prevent the human from navigating a specific course.

Q. What are the legal implications of an AI system having command and control over a vessel which cannot be overridden by the people on board?

---

[15] Changing the educational requirements for seafarer training would require a change to the STCW Code and Convention which needs agreement at the IMO. Developing new regulations can take a considerable amount of time, so an interim solution could be individual nation states defining their own regulations with a risk that the regulations may diverge from international standards as they develop, which could prove to be problematic in the future.

A. Unknown at this stage. This requires further research, and the AAIP has partnered with the University of Swansea Institute of International shipping and Trade Law to start to address the unanswered questions around civil and criminal liabilities of those on board a ship and those in remote control centres (AAIP, 2021).

Regulatory Considerations

There are a number of regulatory requirements that need to be taken into consideration if the technology overrides or does not allow the instructions of the Master.

Seafarer Training, Certification and Watchkeeping (STCW) Code 2015:

- the master has ultimate responsibility for the safety and security of the ship, its passengers, crew and cargo and for the protection of the marine environment against pollution by the ship.

United Nations Convention on the Law of the Sea (UNCLOS) (1982):

- that each ship is in the charge of a master and officers who possess appropriate qualifications, in particular in seamanship, navigation, communications and marine engineering, and that the crew is appropriate in qualification and numbers for the type, size, machinery and equipment of the ship;
- that the master, officers and, to the extent appropriate, the crew are fully conversant with and required to observe the applicable international regulations concerning the safety of life at sea, the prevention of collisions, the prevention, reduction and control of marine pollution, and the maintenance of communications by radio.

International Maritime Organisation Resolution A.893(21) 1999 - Guidelines for voyage planning section 3.4:

- requires each voyage or passage plan as well as details of the plan, should be approved by the ships master prior to the commencement of the voyage or passage.

International Regulations for Preventing Collisions at Sea, 1972

- Nothing in these Rules shall exonerate any vessel, or the owner, master or crew thereof, from the consequences of any neglect to comply with these Rules or of the neglect of any precaution which may be required by the ordinary practice of seamen, or by the special circumstances of the case.

If we introduce an autonomous system that will not accept the instructions of the master or people on board, what are the legal, insurance and ethical implications during that part of the voyage, and how do we demonstrate compliance with the above requirements?

5.4 Information Required

The autonomous function would need to inform the master and people on board why it has taken a specific course of action or offering a specific advisory piece of information. The time required to provide the advice will be dependent on a number of conditions which could include environmental conditions which may range from calm to hurricane force seas, location of the vessel – e.g. deep sea no traffic, in an area with high traffic e.g. Sydney Harbour, or near a coastal area. Time is a significant attribute to safety as the weather or traffic could easily change, so the system would need to be designed for varying conditions of the vessel.

5.5 Assurance

Introducing an AI system as an advisory function is little different to the systems currently installed, if it requires acceptance and implementation of the advice by the human.

Introducing an AI system that is a controlling function would require the safety case for that system to be accepted by the regulatory authorities and this will be extremely challenging as there are currently no regulatory requirements for acceptance of such systems. (Current certification for autonomous vessels generally relies on assuring conventional safety systems, not on assessing the AI-based elements.)

The principles defined in the AMLAS process could be used to develop the system and the body of evidence necessary for the regulatory authorities to satisfy themselves that the system is safe to operate within a defined context of use.

5.6 The likely outcome

In this case the prevention of an accident leading to a loss of life could have been avoided, assuming that systems were autonomous or the Master heeded warnings from an advisory system.

5.7 The practical implications

For autonomous functions to be accepted into the Maritime Sector will require a significant effort in developing an Internationally acceptable assurance framework to allow the vessels to trade in international waters, or an acceptable assurance framework that will allow operation in coastal waters of a single nation state.

The maritime sector would also need to attract and train personnel to work within a safety case environment.

5.8 The Safety-II implications

Introducing an AI system that will not accept commands from a human to prevent an accident from happening could be considered as a way of mitigating the risks. From a safety perspective, how could we ensure that the output of the AI system is correct?

Would we need to put in a protection system that could override the AI system due to a potentially wrong decision, or do we trust the output from the AI sufficiently to negate any override function being installed. As Safety-II focuses on reinforcing "what goes right" and masters are right most of the time, preventing them from taking control – overriding the autonomy – would seem counter to Safety II.

If an AI system was installed on the Costa Concordia and an override was provided, where would this be located, on the vessel or ashore under the final decision of the owner or delegated authority?

6 Conclusions and Further Work

Some common themes can be identified across all the accident case studies:

- The difference between advisory and control autonomous functions. In many sectors such as aviation and maritime there are legal problems with an autonomous function taking control, as the pilot, master or other operator has legal responsibility for safety. This is presumably not such an issue on highways as there are already vehicle control functions in existence, such as automatic emergency braking. Legal changes may have to take place in some sectors.

- We may need to standardize procedures and communications for autonomous functions, i.e. humans may face additional constraints on language and phrasing which may not be well-received.
- Additional authorities – complexities of overrides which may be remote (e.g. for pilot/ground master/shore). Also, there is a possible distinction between direct and indirect stakeholders, i.e. it may be necessary to take into account all those who have real "skin in the game" i.e. may be directly impacted by safety or financial losses, e.g. vehicle owner, insurance company or airfield owner.
- Regulations need to be updated to allow for autonomous functionality: many current regulations were written years ago (especially in sectors such as aviation or maritime) and have evolved slowly over time.
- Overrides may well need to be in place for acceptance of an autonomous system (especially a control system) in many sectors – there will, at least initially, be a lack of trust in new systems. Of course, overrides are often abused and warnings and alarms from systems are switched off to avoid irritation. There will need to be a very low level of 'false-positive-activations' for the systems to be acceptable. On the other hand, the humans involved need to understand what the system will do when activated, not be surprised and misunderstand what the system is doing and incorrectly turn it off. Clearly additional training may be required.
- Transition from partially automated to fully autonomous. The partial case is likely higher risk as someone needs to be ready to take over. A related issue is skills degradation where humans lose the skills necessary to operate the vehicle. Additional training and assessment will be required.
- The complexity of mixed autonomous-enabled and human-operated environments is a big problem because consistent behaviour cannot be guaranteed across all elements (e.g. vehicles on a highway). How will people know how the autonomy will behave across a mixed fleet (e.g. the vision recognition system linked to automatic braking in a vehicle from one manufacturer may recognize different situations to another vehicle)? How would humans 'second-guess' the autonomous systems? This could be impossible unless there is consistent standardization of behaviours.
- The same issues arise where there is communication between autonomous elements. Do these have to trust each other's information? Will there need to be a hierarchy of trustworthiness?
- Gaming AI systems / Spoofing / Hi-jacking of autonomy due to abuse of autonomous function (e.g. for road vehicles stopping if a pedestrian steps out).

- Security – autonomy could be abused or modified remotely for nefarious purposes.
- Cultural / national differences required of AI behaviour. There could be issues in systems which have to cross national / cultural barriers making inappropriate judgements given the changing environments.
- The legal / liability / commercial implications of the AI if an accident does happen. Who is then to blame? Who is liable? If warnings from an autonomous function are ignored who is to blame for an accident? There are also issues for insurance.
- Accident Investigation. There will need to be autonomous-aware, competent accident investigation staff for each sector to establish causes and produce intelligible reports that can be used to reduce future occurrences. This goal may be unrealistic and an "Accident Involving Autonomy" centre may be needed to work across sectors and advise the specific sector accident authorities. This may need nationally-focussed investigations rather than regional bodies for sectors such as highways.
- "Accidental Autonomy" can occur where systems which may be designed and operated with human oversight effectively become autonomous if there is nobody to monitor them or the humans are put in a situation where they cannot effectively monitor the system (intentionally or otherwise). The recent crashes involving the Tesla Autopilot might come into this category, as these systems require constant monitoring and supervision by a driver (despite how they are labelled, e.g. full self-driving).
- A related problem of operator laziness should not be underestimated as if the systems seem to work as they should, why do they need monitoring? There have been many cases of operators or supervisors deliberately or mistakenly not doing all the monitoring they required and autonomous functions may only make this worse. The Uber crash in Tempe Arizona is a case in point (Wikipedia, 2018).
- Many of the proposed autonomous systems require, or require to interact with, a separate control centre or third-party monitoring entity. This could be for operational monitoring (e.g. for breakdown recovery) or could be to enable remote piloting in case of emergency. Remote control adds a whole new level of complexity on top of autonomy and raises questions to do with authority and security.

Two Levels of Autonomy

Perhaps we need to start thinking "out of the box" and ask the question whether some autonomous functionality might be appropriate in most or all emergency situations, where handover to a human operator is impractical. This might be termed "function of last resort", and already exists in automotive with ABS and AEB. It is noted that there is a possible paradigm shift becoming apparent in which, instead of control reverting to a human operator when things go wrong, control of a critical aspect might transfer to this autonomous function.

Hence, we can envisage two levels of autonomy: one which runs day-to-day operations routinely and safely, interacts with operators as necessary and is suitably robust, and a second level which is ready to take over if things go badly wrong and can "get the plane down", or reach an MRC, in severe situations, without any operator intervention.

Conclusion

We hope that the consideration of historical accidents will prompt some further useful thinking about the scope and nature of new autonomous functions.

Acknowledgments Mike Parsons time was funded by the AAIP. Thanks to Alison Hobbs for reviewing.

Disclaimers All views and opinions are those of the authors and not their employers.

References

AAIP (2021), Remote control centres in autonomous shipping, https://www.york.ac.uk/assuring-autonomy/demonstrators/regulation-maritime-remote-control-centres/

AMLAS (2021), Assurance of Machine Learning for use in Autonomous Systems (AMLAS), https://www.york.ac.uk/assuring-autonomy/guidance/amlas/, accessed October 2021

BEA (2021) Serious incident between the Boeing 787-10 registered N16009 and the Airbus A320-214 registered OE-IJF on 20 July 2020 at Paris-Charles de Gaulle airport (Val-d'Oise), Investigation Report, BEA2020-0289

CorwinBrett (2016), 1991 M4 Motorway Crash, https://www.dailymotion.com/video/x4jy8ph, accessed October 2021

DSIWG (2021), SCSC Data Safety Guidance Version 3.3, SCSC, https://scsc.uk/scsc-127F , accessed October 2021

LawCommission (2021), Automated Vehicles, Law Commission https://www.lawcom.gov.uk/project/automated-vehicles , accessed 2021

MIT (2021), Deep learning helps predict traffic crashes before they happen, MIT, https://news.mit.edu/2021/deep-learning-helps-predict-traffic-crashes-1012, accessed October 2021

RAC (2017), Top 10 safety features modern cars should have as standard, RAC, https://www.rac.co.uk/drive/advice/road-safety/top-10-safety-features/ , accessed October 2021

SCSC (2021). Boeing 737 MAX Safe to Fly? - Paul Hampton, Dewi Daniels, SCSC, https://scsc.uk/scsc-162, accessed October 2021

TSB (2016) Transportation Safety Board of Canada, Report A14O0217,
https://www.tsb.gc.ca/eng/rapports-reports/aviation/2014/a14o0217/a14o0217.html

TRL (2020), Safe performance of other activities in conditionally automated vehicles Automated Lane Keeping System, N Kinnear, N Stuttard, D Hynd, S Helman, M Edwards, TRL PPR979, December 2020, https://assets.publishing.service.gov.uk/government/uploads/system/uploads/attachment_data/file/978409/safe-performance-of-other-activities-in-conditionally-automated-vehicles.pdf, accessed November 2021

UNECE 157 (2021), UN Regulation No. 157 - Automated Lane Keeping Systems (ALKS), https://unece.org/transport/documents/2021/03/standards/un-regulation-no-157-automated-lane-keeping-systems-alks, accessed November 2021

Wikipedia (1977), Tenerife Airport Disaster, Wikipedia, https://en.wikipedia.org/wiki/Tenerife_airport_disaster, accessed November 2021

Wikipedia (2018), Death of Elaine Herzberg, Wikipedia, https://en.wikipedia.org/wiki/Death_of_Elaine_Herzberg, accessed November 2021

Wikipedia (2021), 1991 M4 motorway crash, Wikipedia, https://en.wikipedia.org/wiki/1991_M4_motorway_crash, accessed October 2021

YouTube (2019), M4 Motorway Crash – 199, https://www.youtube.com/watch?v=OphNdZhT9Ik, accessed October 2021

A Pipeline of Problems, or Software Development Nirvana? The Challenges of Adopting DevSecOps in a Safety-Critical Environment

Mike Drennan, Paul McKernan, James Sharp

Defence Science and Technology Laboratory (Dstl)

Porton Down, UK

Abstract *The use of DevOps methodologies is now common throughout the technology world, driving a cultural change in the development of software. DevSecOps takes this one-step further, embedding the security measures required in today's hyper-connected world. The adoption of DevSecOps in defence will pose significant challenges to the way that we currently build and deliver software; this will be particularly challenging in the safety-critical domains, such as aviation. In this paper we introduce the principles and technologies proposed in the DevSecOps software development pipeline. An examination into how the US DoD are deploying DevSecOps to maintain operational superiority is given. Two challenges are identified for the UK: providing assurance for safety-critical systems through the DevSecOps pipeline and making the cultural changes necessary to adopt and adapt from the tried and tested methods to this new approach. Finally, we highlight the path that adoption of DevSecOps introduces, identifying the capabilities and further technologies that will naturally be incorporated into this cultural and technological shift for safety-critical software development.*

© Crown copyright (2021), Dstl. This material is licensed under the terms of the Open Government Licence except where otherwise stated. To view this licence, visit http://www.nationalarchives.gov.uk/doc/open-government-licence/version/3 or write to the In-formation Policy Team, The National Archives, Kew, London TW9 4DU, or email: psi@nationalarchives.gsi.gov.uk. DSTL/CP135437.
Published by the Safety-Critical Systems Club. All Rights Reserved.

1 Introduction

The relentless pressure on industry to reduce development cycle times and accelerate delivery has quickly enabled DevOps to establish itself as the enterprise software development practice of choice (Loomba and Wadhwani, 2020). DevOps unifies development and operations with processes and practices that reduce the time from requirement to deployment, whilst also ensuring a high-quality product (Fayollas et al, 2020). Alongside these rapid development procedures is a culture, fostering collaboration amongst all participants in the software life cycle. The DevOps development approach has been adopted by a number of software development teams within industry-leading companies (Microsoft, Facebook, Netflix, Amazon, Etsy etc.) (Altexsoft, 2021).

The ever increasing need to protect against cyber security attacks, whilst still enabling DevOps levels of development velocity, spurred its natural evolution to DevSecOps. The former head of data protection technology at JP Morgan & Chase Co, Omkhar Arasaratnam, stated that "you can't bolt on security in DevOps" (Redwood, 2021). DevSecOps embraces this sentiment by embedding security in every phase of the software life cycle.

The defence industry is acutely aware that its current development approach, applied to increasingly complex software-centric platforms, does not necessarily align with its need to remain operationally superior. As evidenced by the recent announcements from the United States (US) Department of Defense (DoD) (DIB, 2019), the defence industry is looking to the practices of technology companies, such as Google and Netflix, for ways to increase the production rate of high-quality software.

This position paper introduces the terminologies and technologies, and provides initial impressions on, the DevSecOps software development pipeline, as proposed by the US DoD. It considers how these DevSecOps practices will potentially affect the current approaches to Verification and Validation (V&V) and certification. It should be noted that the proposed DevSecOps processes and procedures are not yet fully matured, particularly in relation to developing safety-critical software.

2 Legacy Platform Software Development

Legacy air platform software development has been based upon a conventional V-model lifecycle. This model has traditionally suited safety-critical systems with each stage having V&V criteria developed in parallel to the artefacts being built. More recently, both US and UK defence has adapted to utilise Agile techniques for the development of Mission Systems embedded software development. In addition to prime contractor adoption of Agile, teams are developing software within their own pipelines, with some depots/sub-suppliers now having the capability to deliver directly into the prime contractor's development streams.

The current US/UK military certification approaches, especially when considering safety-critical software, require extensive and coherent evidence that the production processes are robust and the product adequately safe. For software certification, the UK's Def Stan 00-970 (MOD 2020) points to DO-178C (RTCA 2011) as an acceptable means of compliance (AMC); equivalence can be established with the US's MilHbk-516 (DoD 2014) by comparing both standards through their compliance with DO-178C. However, none of the current standards consider the use of DevSecOps (acknowledging that DO-178 is lifecycle agnostic).

Whilst these regulatory requirements would appear to be at odds with the DevSecOps/Agile approach to reducing paperwork, it is possible to see a process that meets the needs of both parties. Perhaps, for example, a process could be established, as an integral part of the development, to push (normally automatically) the relevant artefacts to the regulator as they are produced. The degree of manual intervention and oversight in this process would also need careful consideration.

3 Software Nirvana?

The DoD Enterprise DevSecOps initiative is a joint programme between the Office of the Under Secretary of Defence for Acquisition & Sustainment (OUSD (A&S)), DoD Chief Information Officer, US Air Force (USAF), Defense Information Systems Agency (DISA) and the Military Services. The program aims to provide demonstrable security and quality improvements over the current methods of software development. The initial impetus for the programme came from a Software Acquisition and Practices (SWAP) study conducted by the Defense Innovation Board (DIB) in 2019 (DIB, 2019). The SWAP report recommended a complete overhaul to the way in which the DoD procured, developed, certified and deployed software; describing the current DoD software regulations and processes as "debilitating".

3.1 DevOps

DevOps[1] originated from a need to improve software quality whilst increasing the velocity of development in very competitive time-sensitive markets. The 'Dev' represents those involved in developing software products and services (those pushing rapid change), whilst the 'Ops' includes those who manage and deliver these artefacts (often those preferring stability). Heavily reliant on communication and collaboration, DevOps offers a competitive advantage to organisations, but requires both cultural and process change to marry these 'often opposing' perspectives. The DevOps movement is based on five underpinning values (CALMS):

Culture – shared visions, goals and incentives are enabled through effective teamwork with the minimum of process.

Automation – the utilisation of tools and technologies (Continuous Integration (CI), Continuous Delivery (CD) etc.) will help deliver greater pace, agility, consistency and reliability. Automation is key to many of the benefits of the DevOps approach. Companies such as SpaceX have approximately 200 developers undertaking all of their software development, 95% of which is automated allowing them to concentrate on the difficult 5% (Chaillan, 2019).

Lean – techniques such as mapping value streams, help to identify those activities adding value to a product and also detect sources of waste.

Measurement – metrics help to ensure changes deliver actual improvements. The four key benchmarks (DevOps, 2020) include:
Lead time – period from new requirement to deployment,
Deployment frequency – how often a new release can be deployed,
Failure rate – how often software fails during production,
Mean time to recovery – how long it takes applications to recover from failure once at the production stage.

Sharing – part of the feedback loop in any DevOps environment. A culture of sharing ideas and problems helps an organisation to improve through communal learning.

---

[1] Inspired by John Allspaw and Paul Hammond's talk, "10 Deploys a Day: Dev and Ops Cooperation at Flicker" at the O'Reilly Velocity conference, (Jun 2009) Dubois created his own conference in Belgium called DevOpsDays,(Oct 2009) and the term "DevOps" went viral in the software community.

The key principles ('The Three Ways') of DevOps were introduced by Behr et al. (Behr et al., 2013) in the 'The Phoenix Project', and frame the processes, procedures and practices of DevOps:

The First Way (flow) – understand and increase the flow of work from development to operations.

The Second Way (feedback) – create short feedback loops to enable continuous improvement without substantial rework.

The Third Way (continuous experimentation and learning) – create a culture that promotes the taking of measured risk and learning from failures whilst also understanding that repetition and practice are prerequisites to mastery.

These underlying values and principles, toward delivering higher quality software at pace, form the foundation on which DevSecOps is based.

3.2 DevSecOps

Although the two movements are closely connected through a system of strong values, it would be a simplification to state that the DoD DevSecOps initiative is 'just' DevOps with security built-in (DevOps Institute, 2020). While the early introduction of security is a key DevSecOps differentiator, the DoD initiative includes the tools, network architectures and processes to rapidly develop secure software at the pace of operations. A reference design (DoD CIO, 2019) has been developed to provide guidance on how software teams can best utilise the DevSecOps tools and processes to deliver:

- Applications in a timely and secure manner.
- Functions with security embedded from inception through to operation.
- Reduced accreditation timescales.
- Seamless portability across all environments.
- Continuous Authority To Operate.
- Transformation to a Lean and Agile software development environment.

The key enablers and supporting principles purported to provide these benefits for the DoD include:

Cloud One – A big driver behind the DoD's movement towards DevSecOps was the countless teams it had producing their own software, not all of them successfully, and often not co-located with the end users or each other (Defense Advanced Research Projects Agency (DARPA) in Virginia, Air Force Research

Labs (AFRL) in Ohio, Kessel Run in Massachusetts, Tron in Hawaii etc.). Cloud One provides the network infrastructure for achieving greater interaction through DoD authorised cloud environments (Amazon Web Services (AWS) GovCloud and Microsoft's Azure Government, both with US only access to isolated regions and resources) (AWS, 2020 and Microsoft, 2020). Based on a zero trust architecture model (NIST 2020) Cloud One offers a pay-per-use scalable model for use up to classification Impact Level 6 (secret) (DISA 2018).

Containers – The DoD is driving towards containerising code (from enterprise software to mission systems) into lightweight standalone units that come complete with all dependencies and can run atop an Operating System without confliction. These reusable components can be moved from physical to cloud-based infrastructure and vice-versa. Utilised together with an orchestration tool (Erl 2005) (e.g., Kubernetes) and a Service Mesh[2] (e.g., Istio), applications can be scaled massively (DIB, 2019) and provide capabilities such as encrypted communications, load balancing and self-healing[3].

Centralised Repository – Repo One is a repository that provides a centralised warehouse for source code produced from a plethora of differing projects (be they tools or product code). Software from here can be evaluated and hardened (the process of building in accreditation through the use of security stack sidecar containers: a security process that leverages a zero-trust model that brings in cenetralised logging, utilises whitelisting rather than blacklisting, and is based on behaviour not just signature cyber vulnerability detection) and then digitally signed. These 'hardened' binaries are then stored in the DoD Centralised Artefacts Repository (DCAR) called Iron Bank, providing a pool of easily accessible, secure and reusable components.

Software Factories – A containerised software factory can be instantiated using a set of DevSecOps hardened containers offered in the DCAR. These containers are preconfigured and secured to reduce the certification and accreditation burden and are often available as a predetermined pattern or pipeline[4] (DOD 2019) that needs limited or no configuration. The pipelines can include the toolsets that enable CI and CD techniques, used by some of the most successful companies in the world, to deliver software at pace in a repeatable fashion (Google, 2019). By utilising a fully prepared software development environment, collaborating teams

---

[2] In software architecture, a service mesh is a dedicated infrastructure layer for facilitating service-to-service communications between services or microservices, using a proxy.

[3] Self-healing, in this domain, is the term used to describe any application, service, or system that can: identify that it is not working correctly, and make the necessary changes to restore itself to its correct operational state without human intervention.

[4] A pipeline automates many of the activities in the develop, build, test, release, and deliver phases, and enforces a set of workflows, with minimal human intervention.

The Challenges of Adopting DevSecOps in a Safety-Critical Environment 269

can concentrate effort on building applications. The containerised toolsets help to ensure commonality and interoperability between all contributing organisations.

Continuous Authorisation To Operate (CATO) – DoD systems are required to be approved by an Authorising Officer (AO) against a set of security risk controls before being utilised operationally. Providing the operational context is constantly monitored, once this Authorisation to Operate (ATO) has been signed, its validity endures whilst the process remains intact. Embedding security in the entire application lifecycle could provide a path to CATO and help reduce the long challenging process (NextGov, 2020) of obtaining an ATO.

Platform One – DevSecOps Platform (DSOP) is a collection of hardened containers that are compliant to the Cloud Native Computer Foundation (CNCF) Kubernetes distributions. They contain Infrastructure as Code (IaC) (Wittig 2016) and implement a platform that is compliant with the DoD's DevSecOps reference design (DoD CIO, 2019). Platform One supports DevSecOps by offering managed services to these platforms, these facilities include:

Party Bus – this is a pay per use service that hosts multi-tenant environments (a single software set, with shared licensing etc, supporting multiple tenants, but with each tenant's data being isolated and invisible to other tenants) for test, development and production. These environments provide CI/CD through a host of approved tools (the platform contains an array of tools including 16 programming languages) and capabilities, all benefiting projects wishing to utilise a CATO. The Party Bus is currently being utilised to develop the Advanced Battle Management System (ABMS), which is underpinning the DoD's efforts to interconnect sensors from all the military services into a single network (National Defense, 2020).

Big Bang – also a pay per use/container for larger teams that require a dedicated environment at various classification levels. Again utilising CI/CD pipelines and CATO to build and deliver new hardened project specific containers. Advertised as suitable for projects such as F-35 since tool and environments can be tailored for the user.

Cloud Native Access Point (CNAP) – this service is available on Cloud One and provides access to development, testing and production environments through internet-facing zero trust environments. CNAP allows the use of Virtual Desktop Infrastructure (VDI) on Bring Your Own Devices (BYODs), whist also enforcing device security.

Training – DevOps/DevSecOps are both a culture and a practice step change from traditional software development. Training is essential to understand the principles and includes self-service options together with full on-boarding consultancy (2 month courses to bring your own DevSecOps environment up to full development). The initiative has planned for an annual training capacity of 100,000 personnel.

4 DevSecOps' Adoption in the Current Defence Landscape

Chaillan states that the DoD DevSecOps reference architecture is being designed to have embedded weapon system development capability and has been dismissive of comments suggesting that the methodologies and processes are not a good fit for the high integrity world (Chaillan, 2019).
There is recognition that the real-time safety-critical world may require a different set of tools to that of the other domains, but processes, architectures and culture could be aligned with contemporary cloud, container and service concepts. Some of these concerns are being tackled by DoD DevSecOps subgroups such as:

Continuous ATO Guidance – a team looking at the accreditation and certification of teams, development pipelines and deliverables.

Real-time and Embedded Systems – another team looking at demonstrating Kubernetes can run in a real-time embedded system. Commercial involvement includes Redhat, Wind River, and SpaceX who have already seen some success implementing a Linux based Kubernetes solution on an F-16 legacy system. This Hill Air Force Base team (SoniKube) demonstrated a weapon system solution utilising microservices, but the extent of the Kubernetes clusters' interaction with the aircraft's safety-critical systems (GCN, 2021) is unclear.

In September 2020, a U-2 Dragon Lady flew with mission system software modified to take advantage of containerised Machine Learning (ML) algorithms, orchestrated by Kubernetes (US Air Force, 2021). The successful sortie demonstrated advantages in pooling the hardware resources of the four flight-certified mission/flight computers, without interfering with the aircraft's legacy functionality. The software modifications were carried out by the U-2 Federal Laboratory who are looking to establish and accredit a process for rapid deployment of a technical stack[5] through Air Combat Command.

---

[5] Technical stack refers to the equipment and services required to deliver a technical solution; in this case the delivery of software solutions.

Moreover, some of the largest suppliers of support for embedded safety-critical software systems have started to produce Operating Systems (OS) and infrastructure to enable the implementation of DevSecOps technologies and methodologies. Wind River Systems are currently defined as 'the leader in the Edge computing market' (BusinessWire, 2021), a software philosophy that supports a distributed network placing resources close to the area of need. Wind River have now developed these embedded OS products to include support for Docker containers[6] and Kubernetes orchestration; and have tested real-time applications in a tactical environment by accessing new containers from a repository (a Docker hub) and pulling these into a live operational environment.

Modern commercial and military hardware topologies are already able to support the distributed software architectures that DevSecOps is provisioning. Computing on the Edge[7] 'universal translator' devices (such as General Electric's Remote Interface Units (GE Aviation, 2020) fitted to the F-35 and Boeing airliners), coupled with a high speed bus, could provide the off-the-shelf adaptable interfacing and processing required to host these contemporary software solutions.

4.1 An Anticipated DevSecOps Roadmap

Figure 1 illustrates some of the likely milestones on the road to the digital transformation of software development and support for a typical air platform. Any new strategy will likely aim to utilise a plethora of contemporary software development techniques, including the use of Artificial Intelligence (AI) and ML, where appropriate, to provide warfighting advantage; much of which will borrow processes and methodologies from successful commercial enterprise. This transformation will undoubtedly run in parallel with a programme of cost reduction through rationalisation and competition.

---

[6] A Docker container image is a lightweight, standalone, executable package of software that includes everything needed to run an application: code, runtime, system tools, system libraries and settings.(https://www.docker.com/resources/what-container).

[7] Edge computing is an approach to computing that places the computing capacity and data storage close to the data source, decentralising computation and reducing networking bandwidth requirements.

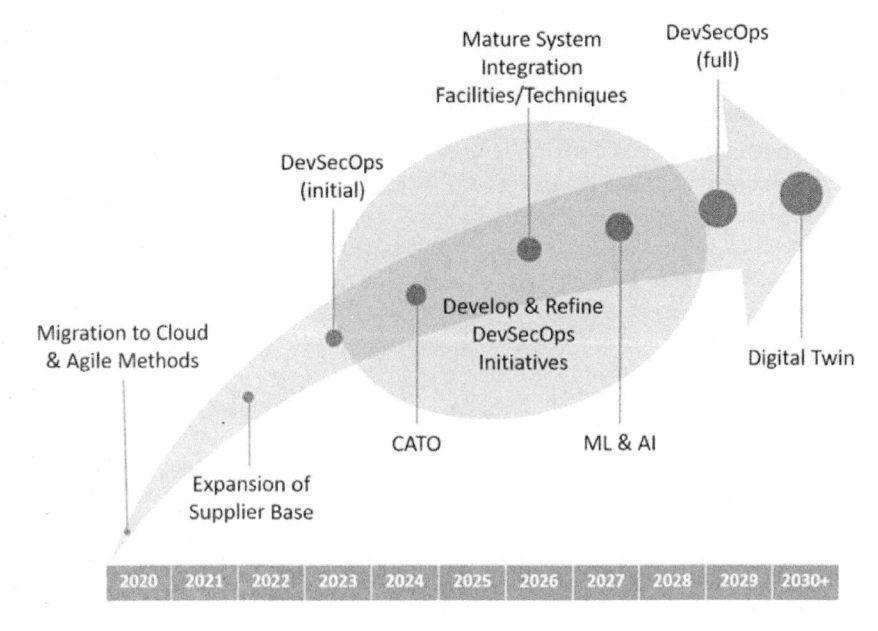

Fig. 1. Expected Contemporary DoD Software Development Trajectory

Even if DevSecOps can 'embed' some of the application security, there is still pressure on the certification processes to match this new pace of development. In addition to fielding new capabilities, the ability to counter emerging Electronic Warfare (EW) and cyber vulnerabilities are also time critical in frontline environments. There is a requirement to get enough velocity into these safety processes to make any new development worthwhile. Challenges include:

Automatic Evidence Generation – if we are to regulate at pace we may have to remove the person-in-the-loop. A process will have to be cultivated for automatically extracting airworthiness certification evidence from a rapid software development pipeline. Tools for mining this evidential data will have to be qualified and be adept enough to check for relevance and completeness.

Regulator Engagement – Given the novel approach of DevSecOps' software development it is essential to have early engagement with the regulatory authorities on finding paths to certification. Liaison with the certification authorities will be vital to ensure the outputs from these new processes provide the evidential artefacts required from the regulators.

Test and Evaluation (T&E) – aircraft T&E activities will also require updating to integrate into any new lean qualification lifecycle. These processes will likely require a greater use of automation, simulation and models. Greater emphasis on models and simulations may introduce synchronisation and validation challenges to avoid behavioural divergences between these digital representations and reality. Whilst 'enterprise' solutions may be more suited towards greater autonomy in testing, other system applications may require more traditional processes (independence, testing variety etc.). Not even the greatest DevSecOps advocate and the driving force behind the reference design (Nicholas Chaillan, USAF Chief Software Officer) suggests that all testing for all systems will be automatic (Chaillan, 2019). Although T&E discussions with the Air Force and nuclear safety organisations are ongoing there is still a host of unanswered questions, from the embedded safety-critical community, on how DevSecOps can work for them (Chief Software Office, 2021).

4.2 The Impact of Adopting DevSecOps on an Existing Defense Platform

With in-service air platforms already relying heavily on complex software, none more so perhaps than Lockheed Martin's (LM) F-35, which has been described as 'a flying software programme', the software life cycle of some in-service platforms have been considered 'ripe' for improvement. Additionally, with Agile and DevOps now being taught at graduate level this transformation of defence software development may help to attract new talent into the defence industries.

The military have long recognised that the technical agility of their air platforms is dependent on their ability to produce the software to integrate new capabilities. Thus for military equipment (not just air platforms) software development is an area where the new Agile and DevSecOps paradigms could deliver significant benefits. The DoD has already noted that they are in a race with other nations and losing the race could mean surrendering the domain to their opponents; it is worth noting at this point that just producing software faster is not the answer, the entire enterprise from contracting, through requirements development, coding, testing and certification needs to provide a fast and effective process for delivering capability.

In response to the DIB SWAP study (DIB 2019) the F-35 Project Executive Officer identified the F-35 program as a pilot for software acquisition modernisation (Lockheed Martin, 2019). To this end, the F-35 programme is introducing new tools and methodologies from the DevSecOps initiative, which will have a significant impact on the F-35 software development ecosystem. Likely changes include:

New depots – the use of containerised software factories and cloud connectivity will enable development away from the traditional LM sites. This capability supports the DoD requirement to contract from best in class and will help to stimulate competition. The fact that the DoD cloud infrastructure will now be used to develop and store elements of our national software assets may be an issue for the UK Government.

Expansion of development teams – the enhanced ability to deploy makes integrating new producers simpler and more cost effective. These new producers will have to achieve compliance with continuous assurance audits before being permitted to feed safety-critical software into the LM main software stream.

UK development – wrapping up the Software Development Environment (SDE) in a container with all the necessary tools and dependencies could be an opportunity for Partner Nations to develop their own sovereign software loads. This will depend on the access non-US partners get to the DoD clouds (currently only US based admission) (TripWire, 2020). Even if foreign software teams can participate, it may require enormous investment to provide the necessary infrastructure and accreditation to start developing software modifications.

Containerised functionality – the proposed new software architectures allow for the execution of applications in multiple distributed containers. The container orchestrator manages the containers to provide services (functionality) that can be easily scaled and offer built-in redundancy. These distributed architectures are traditionally suited to the enterprise software domain where the services are reused and grouped together to form multiple business applications. The dispersed applications rely heavily on microservices that provide control and communication functionality. The routing of requests and data amongst the network of services can introduce indeterminate execution times (Sampaio et al., 2019). These temporal issues add to the complexity of the safety-critical software arena and will require novel approaches to provide the necessary certification evidence.

For the F-35, LM is unlikely to change its current safety-critical software development process to methodologies and architectures with unknown risks; but may, however, start to develop new applications utilising the DoD prescribed techniques. Thus, it should be noted at this point, that whilst this approach may provide a working solution, the continued use of existing V&V certification processes make it unlikely that the full benefits of a DevSecOps environment will be realised.

Pace of development – the use of CI/CD will inevitably lead to delivery at pace with faster iteration rates. To keep up with these rapid development cycles security assurance will need to adapt through the use of CATO[8]. To provide timely capabilities, the challenge of providing software certification, of the product, in

---

[8] Where CATO ensures that the processes remain compliant with the relevant standards.

this new enterprise must be solved and although the subject of various DevSecOps working groups; an exemplar solution has yet to emerge.

5 A Pipeline of Problems?

Most recently Will Roper, the lead author of the DIB SWAP study has moved (Defence News, 2021) to support the UK RAF's aspirations to modernise the way that safety and mission critical software is developed, enabling an alignment with the US DoD's approaches to date. Such an alignment should look to ensure that any relevant industry initiatives, which should be welcomed and regarded as essential, need to understand and co-ordinate with other UK efforts.

US progression in the DevSecOps world will directly influence not only UK F-35 development, V&V and its certification, but provide learning in support of the UK's next generation warfighters. Thus, the US direction must be conveyed to UK stakeholders and assurances sought on access to information likely to provide productive engagement with the UK regulator, not just to support the continued certification of F-35, but also to provide commonality where appropriate, and to leverage learning for the UK.

At the heart of 'high integrity' DevSecOps is a clash between the depth and breadth of engineering excellence required to produce safe software and the delivery-focussed drive of these new development methods. Successfully understanding and addressing this conflict of speed and the greater engineering requirements of safety-critical software will be key to the UK adoption of this methodology. One of the key enablers to this work will be the development of sufficient UK SMEs to provide intelligent customer support.

As stated earlier, producing software faster loses most of the benefits if the product cannot be certified for use at a similar pace. Expecting or demanding the regulator change to meet the new challenge is a fraught path. More realistically, the new processes must be developed to provide the evidence, much of it produced automatically, needed to enable certification using existing approaches. However, early and constant engagement with the regulator should encourage faster certification and eventually support the regulator in moving to a stance that matches certification to the speed of delivery. Thus, as this new landscape evolves, further analysis will be required to help provide recommendations on supporting the likely adoption of these techniques in both current and future systems, ensuring that the safety-critical software remains safe and conformant to appropriate guidance and standards. Nicolas Chaillan (Air Force CSO) was given the authority to fundamentally change the DoD's software engineering culture and practice (DIB, 2019). Embracing the DoD DIB's SWAP recommendations and aligning them with his expertise in Cloud computing and Cybersecurity, he directed and championed the benefits of the DevSecOps initiative. Committed to

DevSecOps, Chaillan has stated that teams producing software outside of a DevSecOps environment "is borderline criminal" (Chaillan, 2019) and "we should not be in a situation where we are not using these principals".

However, during the final throws of putting this paper together Nicholas Chaillan announced his departure as Chief Software Officer for the USAF (Defence News, 2021a). Throughout his 3 year tenure Chaillan has been at the forefront of the introduction and use of DevSecOps methodologies and processes within the USAF and wider DoD. He has been responsible for the adoption of rapid development techniques and streamlining deployment processes. But, Chaillan has been frustrated with the leadership of the DoD for some time, stating their lack of alignment with his departments DevSecOps vision as a major contribution for his decision to resign. Without Chaillan leading the USAF's software modernisation programme there is a real risk of stagnation and a half-hearted attempt at transformation.

As the UK seeks to implement our own DevSecOps environment there is a risk that we also find this a challenge, perhaps more so as we do not appear to have identified a UK champion to lead the endeavour. It should be noted that, whilst a technically gifted software expert is required to provide the vision, this individual needs equivalent support from people with contracting and acquisition expertise. Without this coherent approach any UK initiative is likely to flounder.

There is a further challenge to UK adoption as the US process allows for the rapid addition and removal of suppliers. This approach, whilst challenging in the US environment, runs counter to the UK approach of having a prime and a stable team of sub-suppliers. For DevSecOps to work in the UK an acquisition model that meets both MOD and UK Industry aspirations will have to be developed. For some of the most recent platforms, the US aspiration is to deliver software updates at regular monthly intervals and, to ensure coherence within the programme, the UK needs to adopt this accelerated pace. The current pace of change is already presenting a major resourcing challenge to the MOD, and the move to development at a higher 'DevSecOps' velocity (without changes to our current certification processes) will inevitably exacerbate this challenge. However, the latest evidence from the US seems to suggest that a 6 monthly delivery cadence may not be required by the operators nor deliverable with current resources. Thus the decision on delivery schedules remains the subject of debate, and should eventually be driven by operator need.

6 The Road Ahead

Much of the DevSecOps methodology and process has been derived from the 'enterprise' software world, which employs applications and technologies to support commercial needs and where working at scale with redundancy are key business enablers. Some of these techniques do not usually sit well in safety-critical software development, but Chaillan's DevSecOps mandate is already starting to influence air platform programmes.

An integrated UK F-35 software development house would require considerable investment (networking infrastructure, secure cloud access, tools and training etc.) but would provide both immediate benefits to the UK F-35 programme (industrial access to F-35 technologies, greater influence on UK capabilities etc.) and would allow UK industry to prepare for future advanced software development ecosystems such as TEMPEST[9].

Modern weapon platforms will require the deployment of software updates and fixes at speed to maintain their combat edge. This change in the velocity of software development will enable operational advantage through:

- Reduction in time from cyber vulnerability to fix
- Quick responses to changing Electronic Warfare (EW) threats
- Rapid updates to vehicle and weapon delivery capabilities

It is clear therefore, that the UK has to be able to mesh velocity with integrity: DevSecOps will be at the centre of F-35 software development, and could be an opportunity to bring positive affects to the whole of the UK air platform certification programme. The UK's regulator for defence aviation, the Military Aviation Authority (MAA), is aware of these wholesale development lifecycle changes to air platforms under its jurisdiction. Whilst the MAA has begun to engage with those proposing and delivering these novel and rapid techniques, it is likely that considerable investment, from both the certification agency and suppliers, will be required to mesh their often disparate needs.

Continuing airworthiness certification is a key enabler towards the goal of rapid capability insertion, but it requires evidence that a system has not been adversely affected by modification. Increasing the pace of change often increases the risk of introducing faults with these alterations. There is a need to provide the required evidence out of a lean and automated software deployment pipeline. This certification documentation should ideally be generated automatically as part of the development process, and amalgamated into artefacts suitable for audit by the regulatory authorities. Some tough questions remain unanswered for the F-35 programme:

---

[9] TEMPEST is the UK programme aimed at delivering a next generation combat aircraft (https://www.raf.mod.uk/what-we-do/team-tempest/).

- Will the UK have access to the new DevSecOps processes that will be providing future certification evidence?
- Will the UK have access to the certification evidence generated by these new processes?
- Can we adapt these processes, or the certification path, to make use of these rapid capabilities?

Moreover, for the future of UK Defence, and in the beneficial adoption of DevSecOps, it is clear that, with software as the focal point of future platforms, there will be a requirement for continuous funding to support CI/CD. This financial support will have to be enduring to allow continuous software updates. If we want to integrate software from a plethora of functionalities (Mission, Cyber, EW, Information Architecture (IA), Safety etc.), then we also need to integrate their funding streams. From inception to out-of-service, the support for the software elements of mission, safety and information assurance, from development to certification, must be considered as a whole and funded as such.

References

Altexsoft (2021) DevOps: Principles, Practices and DevOps Engineer Role https://www.altexsoft.com/blog/engineering/devops-principles-practices-and-devops-engineer-role/. Accessed 20 September 2021

AWS (2020) GovCloud Introduction https://aws.amazon.com/govcloud-us/?whats-new-ess.sort-by=item.additionalFields.postDateTime&whats-new-ess.sort-order=desc. Accessed 3 November 2020

Behr K, Kim G and Spafford G (2013) The Phoenix Project: A Novel About IT, DevOps, and Helping Your Business Win, First Edition, IT Revolution Press

BusinessWire (2021) Wind River Recognized as #1 in Edge Compute OS Platforms https://uk.finance.yahoo.com/news/wind-river-recognized-1-edge-130000471.html. Accessed 7 January 2021

Chaillan N (2019) DevSecOps for DoD presentation to the Defence Acquisition University, https://media.dau.edu/media/Nicolas+Chaillan+Presents+DevSecOps+for+DoD+%28DSOD%29/1_pv3s5nxk/62956591.

Chief Software Office (2021) DevSecOps FAQs, Assistant Secretary of Acquisition https://software.af.mil/dsop/frequently-asked-questions-faq/. Accessed 3 January 2021

DIB (2019) Software Is Never Done: Refactoring the Acquisition Code for Competitive Advantage, Defense Innovation Board, 3 May 2019

DISA (2018) DISA Cyber Standards Branch (RE11) Cloud Computing Security Requirements Guide May 2018 https://disa.mil/-/media/Files/DISA/News/Events/Symposium/Cloud-Computing-Security-Requirements-Guide

Defence News (2021) Former US Air Force acquisition czar could help the UK build its future fighter https://www.defensenews.com/digital-show-dailies/dsei/2021/09/14/former-us-air-force-acquisition-czar-could-help-the-uk-build-its-future-fighter/. Accessed 29 September 2021

Defence News (2021a) Air Force chief software officer knocks DoD as he departs https://www.defensenews.com/battlefield-tech/it-networks/2021/09/02/air-force-chief-software-offer-knocks-dod-as-he-departs/. Accessed 8 October 2021

DevOps (2020) Measuring DevOps Performance, https://devops.com/measuring-devops-performance/. Accessed 28 October 2020

DevOps Institute (2020) DevSecOps and ITIL4 (Agile, DevSecOps and Cybersecurity) https://devopsinstitute.com/devsecops-and-itil4/. Accessed 27 October 2020

DoD Chief Information Officer (CIO) (2019) DoD Enterprise DevSecOps Reference Design, Version 1.0 https://dodcio.defense.gov/Portals/0/Documents/DoD%20Enterprise%20DevSecOps%20Reference%20Design%20v1.0_Public%20Release.pdf?ver=2019-09-26-115824-583

DoD (2014) MIL-HDBK-516C Department of Defense Handbook Airworthiness Certification Criteria, December 2014

Edwards AFB (2021) Team Edwards helps pave way for new F-35 ODIN hardware https://www.edwards.af.mil/News/Article/2390143/team-edwards-helps-pave-way-for-new-f-35-odin-hardware/. Accessed 7 January 2021

Thomas Erl (2005) Service-Oriented Architecture: Concepts, Technology & De-sign. Prentice Hall, ISBN 0-13-185858-0

Fayollas C, Bonnin H, Flebus O (2020) SafeOps: a concept of continuous safety, 16th European Dependable Computing Conference (EDCC), IEEE

GCN (2021) Why the Air Force put Kubernetes in an F-16 https://gcn.com/articles/2020/01/07/af-kubernetes-f16.aspx. Accessed 3 January 2021

GE Aviation (2020) Remote Interface Unit (RIU-303) https://www.geaviation.com/sites/default/files/remote-interface-unit-RIU-303-datasheet.pdf. Accessed 14 December 2020

Google (2019) Accelerate State of DevOps Report 2019 https://services.google.com/fh/files/misc/state-of-devops-2019.pdf. Accessed 13 November 2020

Government Accountability Office (GAO) (2021) F-35 Sustainment, U.S https://www.gao.gov/products/GAO-20-665T. Accessed 8 January 2021

Lockheed Martin (2019) F-35 Software Modernisation Summit Working Group, LM Global Visitor Center, Arlington VA

Loomba S and Wadhwani P (2020) DevOps Market Size by Component Report, Global Market Insights https://gminsights.com/industry-analysis/devops-market. Accessed 20 September 2021

Microsoft (2020) Azure Government https://azure.microsoft.com/en-us/global-infrastructure/government/how-to-buy/. Accessed 3 November 2020

Microsoft (2021) DevOps vs. Agile https://azure.microsoft.com/en-gb/overview/devops-vs-agile/. Accessed 20 September 2021

MOD (2020) UK Ministry of Defence Standard 00-970 Design and Airworthiness Requirements for Service Aircraft Part 1, Issue 17, August 2020.

National Defense (2020) Advanced Battle Management System faces headwinds https://www.nationaldefensemagazine.org/articles/2020/9/11/advanced-battle-management. Accessed 13 November 2020

NextGov (2020) Shift Left: DevSecOps and the Path to Continuous Authority to Operate https://www.nextgov.com/ideas/2020/07/shift-left-devsecops-and-path-continuous-authority-operate/167223/. Accessed 13 November 2020

Nicolas Chaillan (2020) DevSecOps for DoD https://cdnapisec.kaltura.com/index.php/extwidget/preview/partner_id/2203981/uiconf_id/39997971/entry_id/1_pv3s5nxk/embed/dynamic. Accessed 15 December 2020

Nicolas Chaillan (2021) Air Force Chief Software Officer, Assistant Secretary of Acquisition, Chief Software Office https://software.af.mil/team/nicolas-m-chaillan-hqe/. Accessed 8 January 2021

NIST (2020) NIST Special Publication 800-207 Zero Trust Architecture August 2020 https://doi.org/10.6028/NIST.SP.800-207

Redwood S (2021) https://myredfort.com. Retrieved 20 September 2021

RTCA (2011) DO-178C: Software Considerations in Airborne Systems and Equipment Certification, December 2011

Sampaio A, Rubin J, Beschastnikh I, Rosa N (2019) Improving microservice-based applications with runtime placement adaptation, Journal of Internet Services and Applications, Springer Open

The Basics (2019) QA Training Certified DevOps Foundation Course (DOIFOUND), The DevOps Institute & QA

TripWire (2020) What is Amazon GovCloud https://www.tripwire.com/state-of-security/security-data-protection/cloud/what-is-amazon-govcloud/. Accessed 14 December 2020

US Air Force (2021) U-2 Federal Lab achieves flight with Kubernetes https://www.af.mil/News/Article-Display/Article/2375297/u-2-federal-lab-achieves-flight-with-kubernetes/. Accessed 3 January 2021

Wittig A, Wittig M (2016) Amazon Web Services in Action. Manning Press. p. 93. ISBN 978-1-61729-288-0

Disclaimer This article is an overview of UK MOD sponsored research and is released for information purposes only. The contents of this article should not be interpreted as representing the views of the UK MOD, nor should it be assumed that they reflect any current or future UK MOD policy. The information contained in this article cannot supersede any statutory or contractual requirements or liabilities and is offered without prejudice or commitment.

Formal verification of railway interlocking and its safety case

Alexei Iliasov

    The Formal Route

Dominic Taylor

    Systra Scott Lister

Linas Laibinis

    Vilnius University

Alexander Romanovsky[1]

    The Formal Route and Newcastle University

Abstract *The increasing complexity of modern interlocking poses a major challenge to ensuring railway safety. This calls for application of formal methods for assurance and verification of their safety. We have developed an industry-strength toolset, called SafeCap, for formal verification of interlockings. Our aim was to overcome the main barriers in deploying formal methods in industry. The approach proposed verifies interlocking data developed by signalling engineers in the ways they are designed by industry. It ensures fully automated verification of safety properties using the state-of-the-art techniques (automated theorem provers and solvers) and provides diagnostics in terms of the notations used by engineers. In the last two years SafeCap has been successfully used to verify 26 real-world mainline interlockings, developed by different suppliers and design offices. SafeCap is currently used in an advisory capacity, supplementing manual checking and testing processes by providing an additional level of verification and enabling earlier identification of errors. We are now developing a safety case to support its use as an alternative to some of these activities.*

---

[1] Contact author's email: alexander.romanovsky@formal-route.com

© (Iliasov, Taylor, Laibinis, Romanovsky) 2022.
Published by the Safety-Critical Systems Club. All Rights Reserved.

1 Railway Signalling

Effective signalling is essential to the safe and efficient operation of any railway network. Whether by mechanical semaphores, coloured lights or electronic messages, signalling allows trains to move only when it is safe for them to do so. Signalling locks moveable infrastructure, such as the points that form railway junctions, before trains travel over it. Furthermore, signalling often actively prevents trains travelling further or faster than is safe.

There are two main safety principles shared by all signalling systems:

- A schema must be free from collisions. A collision happens when a train occupies the same physical space as another train or (at a level crossing) a road vehicle. Signalling systems uphold this principle through the use of signalling routes, and block sections.
- A schema must be free from derailments. A derailment may happen when a set of points moves underneath a train. To avoid this, a point must be positively confirmed to be locked in position before a train may travel over it and held in that position as a train does so.

At the heart of any signalling system there are one or more interlockings. These devices constrain authorisation of train movements as well as movements of the infrastructure to prevent unsafe situations arising. One of the earliest forms of computer-based interlocking was the Solid State Interlocking (SSI) (Cribbens 1987), developed in the UK in the 1980s through an agreement between British Rail and two signalling supply companies, Westinghouse and GEC General Signal. SSI is the predominant technology used for computer-based interlockings on UK mainline railways. It also has applications overseas, including in India, Australia, New Zealand, France and Belgium. Running on bespoke hardware, SSI software consists of a core application (common to all signalling schemes) and site-specific geographic data. SSI GDL (Geographic Data Language) data configures a signalling area by defining site specific rules, concerning the signalling equipment as well as internal latches and timers that the interlocking must obey. Despite being referred to as data, an SSI GDL configuration resembles a program in a procedural programming language and, as such, is referred to as a program in this paper. Such a configuration is iteratively executed in a typical loop controlling the signalling equipment.

The increasing complexity of modern digital interlockings, both in terms of their geographical coverage and that of their functionality, poses a major challenge to ensuring railway safety. This calls for application of rigorous methods, in particular formal methods, for assurance and verification of safety and other crucial properties of such systems. Even though formal methods have been successfully used in the railway domain (e.g. (Badeau and Amelot 2005)), their industry application is scarce, typically focusing on relatively simple interlockings used in metros and other urban lines. In spite of a large body of academic studies addressing issues of formal verification of railway signalling systems, they usually remain an academic exercise due to a prohibitive cost of initial investment into their industrial deployment as well as the inherent limitations of the formal techniques used (such as the state explosion for model checking techniques or substantial manual efforts for applying provers). There are a number of reasons for this;

- signalling engineers need to learn mathematical notations and formal reasoning to effectively apply them;
- many verification techniques and the supporting tools proposed cannot be applied to analysing real modern interlockings due to their poor scalability;
- companies need to drastically change the existing development processes in order to use them;
- the development of formal techniques and tools in academia is seldom driven by the chief aim of deploying them in industry.

The paper first introduces a novel tool-based approach that addresses the above issues by

- verifying the signalling programs and layouts (schemas) designed by signalling engineers in the ways they are developed by industry,
- ensuring fully-automated and scalable verification of safety properties using the state of the art verification techniques (in particular, automated theorem provers and solvers), and
- providing diagnostics in terms of the notations used by the engineers.

All together, this ensures that the developed method and tool can be easily deployed to augment the existing industrial processes of developing complex *mainline* interlockings in order to provide extra guarantees of railway safety.

The remaining part of the paper discusses the proposed verification method, its industrial deployment and application for safety verification of a substantial number of live projects conducted by major signalling companies in the UK, as well as our ongoing work towards developing a safety case to allow the approach to be deployed as a (partial) replacement for the manual checks and testing/simulation widely used now for safety assurance by the railway industry.

Our earlier paper (Iliasov et al 2018) discusses a prototype (experimental) version of the method and the tool, provides additional information on how the SSI verification is conducted and discusses a small case study. Since this earlier paper was written the tool has been substantially reworked and enhanced to improve its usability, scalability, reliability, and the quality of reporting, and to extend the verification coverage. During this period the tool has been demonstrated, validated, and proven in an operational environment, and approved by the cross-industry SSI-Applications Group led by Network Rail for use in UK signalling projects for automated railway signalling verification. All this has allowed us to deploy the approach on multiple commercial railway signalling projects in the UK.

2 SafeCap for Solid State Interlocking

The SafeCap platform (Iliasov et al 2013, Iliasov et al 2014) is a general open extendable Eclipse-based toolkit (Des Rivieres and Wiegand 2004) for modelling railway capacity and verifying railway network safety developed in a number of public projects led by Newcastle University. The platform's main purposes have been to support academic research and to help in exploring how new ideas could be efficiently used by the railway industry. It allows the users to design stations and junctions relying on the provided domain specific language (SafeCap DSL - (Iliasov and Romanovsky 2012)) and to check their safety properties, simulate train runs as well as evaluate potential improvements of railway capacity by using a combination of theorem proving, SMT solving and model checking.

The SafeCap DSL allows the designers to rigorously and unambiguously define a model of the given railway network (e.g., stations or junctions) by providing a formal, graph-oriented way of capturing railway schemas and some aspects of signalling. Various concepts of a railway schema such as signals and signalling solutions, speed limits, stopping points and so on can be incorporated via DSL extension plug-ins. Such plug-ins introduce new data (as custom annotations) and the supporting logic (as additional logical constraints or relationships). Such a tool architecture allows us not to commit to any regional technology and thus to offer a broadly similar approach for a range of legacy and current technologies.

The SafeCap verification and proof back-ends enable automated reasoning about static and dynamic properties of railways or their signalling data. In the course of platform development, it has been substantially extended by adding new simulators, state-of-the-art solvers and provers, as well as the support for importing the existing designs in a wide range of signalling frameworks supported by industry.

In the last 5 years, initially using the SafeCap platform as the experimental prototype, a team of computer scientists and signalling engineers has been working on developing a targeted SafeCap-based industry-strength toolset for formal verification in the course of the SSI development projects. Our aim from the outset was to overcome the typical barriers in deploying formal methods in industry. In the course of this work we have removed the SafeCap functionalities that are not required for SSI verification (such as animation, simulation, and capacity measurement), simplified the tool architecture, and replaced all its solvers and provers with a dedicated inference-based symbolic prover that outperforms all known state-of-the-art provers when used on the railway models that our SSI-targeting tool needs to verify. The core requirements set for this work were that the method and the tool:

- use the existing industrial signalling notations for the inputs and for reporting the results of verification;
- fully hide formal methods from the engineers that use them;
- provide fully automated verification;
- are scalable to allow full safety verification of any existing UK interlocking within a few minutes.

As we explained earlier an SSI GDL configuration resembles a program in a procedural programming language. An SSI system is a continuously running control system that executes a global control loop composed of three stages:

1. polling of inputs (the current states of train detection, points, signals and so on);
2. computation of necessary responses, as well as,
3. formation and transmission of equipment control commands.

The input and output stages are generic and their safety argument is provided once for a particular underlying hardware implementation. SafeCap verification of system safety is concerned with the middle stage – the response computation; the safety of other stages is assured through the safety cases underpinning the generic interlocking hardware and the site-specific wiring connecting that hardware to trackside equipment. This stage is unique to every geographic area and explicitly refers to the equipment drawn on the area *scheme plan* – a diagrammatic depiction of a railway.

The central question in the verification of signalling correctness is what constitutes a safe signalling design. Certain basic principles are universally accepted, for instance, the absence of train collisions and derailment. However, it is almost hopeless to verify the absence of such hazards in the strictest possible sense, not least because there are many real-world limitations:

- trains are driven in accordance with signals, but signals can only convey very limited information, which is only readable for a short distance on the approach to each signal;
- trains have different braking characteristics and must all be given sufficient warning of the need to stop before a red signal;
- drivers occasionally misjudge braking and pass red signals;
- there are some situations that cannot be fully protected by a signalling system, such as allowing trains to couple together or share a platform, for which the basic principles are upheld through human competence and operational procedures.

Instead, correctness is established not against the basic principles but rather against *signalling principles* derived from the basic principles and designed to enable railway operation with an acceptable level of risk and failures: the conditions under which trains can be authorised to move by signals; the indications ('aspects') displayed by those signals and their meanings; when points can move; provision of 'overlaps' beyond red signals in which a train can stop safely if the driver misjudges braking; etc.. Such signalling principles are carefully designed by domain experts and documented in standards such as (Network Rail 2015). However, they can vary between regions and do change over time. At the time of writing this paper the SSI SafeCap verifies 55 signalling principles.

For the purposes of verification, each signalling principle to be ensured is rendered as an *inductive safety invariant* – a system property that must hold when a system boots up and must be maintained (or, equivalently, be re-established) after any state update.

One example of a SafeCap safety invariant is an invariant which checks that whenever the commanded position of a set of points is changed (triggering the points to move) all train detection sections over the points are proved clear:

```
forall p:Node
 point_c(p) != "point_c'p"(p) =>
  "pointclear-
tracks"["Node.base"~[{"Node.base"(p)}]]
        /\ track_o == {}
```

As with all safety invariants, before this formal notation is developed we define and agree on the semi-formal formulation of the invariant:

```
[for]
 every point commanded to a new state
[it holds that]
 all point tracks are proved clear
```

Another example is an invariant that checks that on setting a route, all the route sub routes are locked:

```
forall r:Route
 r:cover(route_s \ "route_s'p") =>
   (forall sr:SubRoute
    sr : "route:subrouteset"[{r}] /\ "SubRoute.ixl" =>
           sr : subroute_l
        )
```

This is the semi-formal notation of the invariant:

```
[for]
every route being set
[it holds that]
all the interlocking-contained sub routes of
                            the route are locked
```

Verification of an inductive safety invariant is understood as the problem of checking that any safety invariant is respected by every state update. Technically this is done be generating conjectures (also called proof obligations – POs) of the form "*if an invariant holds in a previous state and a certain state update happens, is it true that the invariant holds for the new state?*". Formally, a conjecture is represented as a logical sequent consisting of a number of hypotheses (H) and a goal (G), denoted as $H \mid\text{-} G$.

The number of such conjectures is $m * n$, where m is the number of safety invariants (circa 50) and n is the number of possible state updates (typically around 10,000). This is a small number when contrasted against the number of potentially reachable states (circa $2^{2,000}$). The complexity measured in the number of conjectures grows linearly with the system size. In practice, however, it is possible to automatically prove or disprove the vast majority of conjectures with only a tiny number (less than 1 in 10,000 for recent results) of provable conjectures failing to prove and resulting in a false positive. Achieving such a level of proof automation is not easy and requires a number of customised techniques developed and integrated into SafeCap.

The major steps of SafeCap verification process are depicted in figure 1:

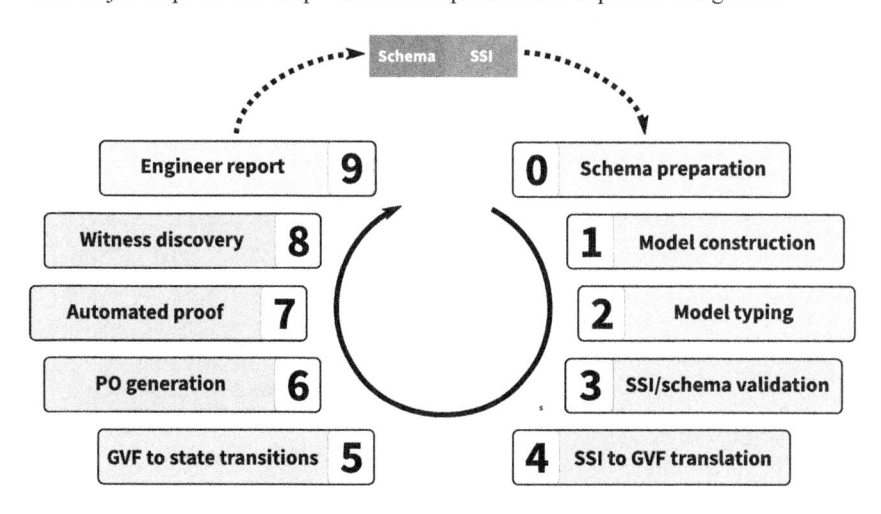

Fig. 1. SafeCap verification process

There are four major stages: preparation of input (steps 0-3), translation and symbolic execution of the signalling logic (steps 4-5), formal verification (steps 6-8), and report generation (step 9). All but the very first stage are automated. In the first stage, input preparation turns an electronic image of a scheme plan into a mathematical model. In the next stage, the obtained model is translated into a large number of individually simple state transitions. In the third stage, formal verification is employed to check that these state transitions are safe. In the last stage, the engineer report is generated to present findings of safety violations in the form understandable to engineers.

Figure 2 depicts the overall data flow of the approach and cross-links it with the steps in figure 1. The corresponding data transformation steps (arrows) are annotated by numbers referring to the respectively numbered steps of figure 1. Intuitively, the left branch focuses more on the system statics (railway scheme, its various constituent elements and their relationships), while the right one deals with the system dynamics related to SSI signalling. The middle branch covers industrial standards, formulating safety principles relevant to the scope and operation level of SSI and then deriving formal safety invariants. Note that the solid arrows in figure 2 depict automated actions by the SSI SafeCap tool, while dashed ones are manual activities.

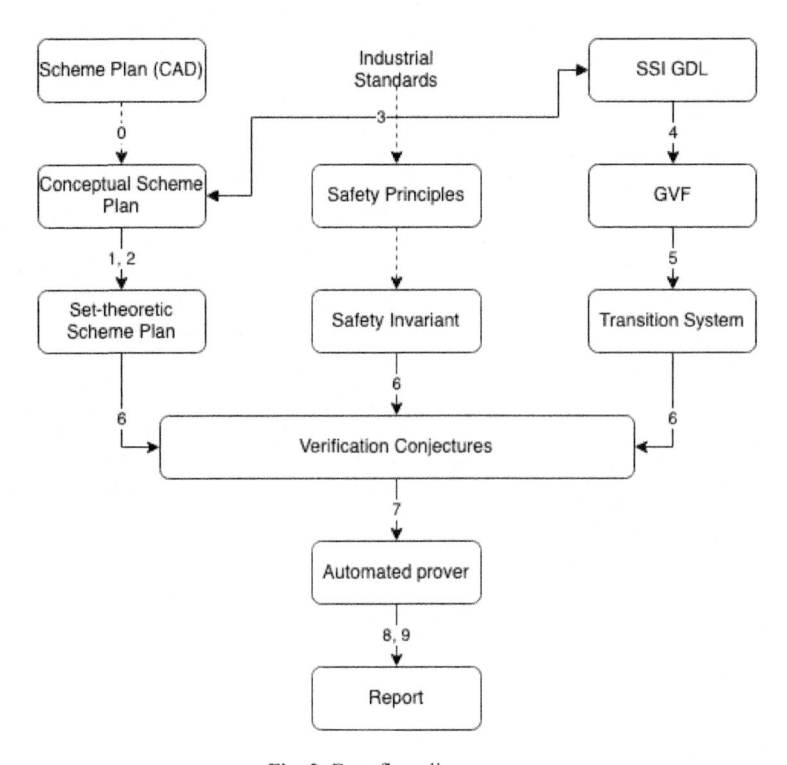

Fig. 2. Data flow diagram

In general, each subsequent layer downwards adds more rigour and formality. The data representations of the third layer (i.e., set theoretical scheme plan, safety invariant, and transition system) are based on the same underlying mathematical language and can be considered as parts of the overall formal system model ready to be verified.

At the top level, the inputs defining a particular area are a relevant scheme plan and the associated SSI GDL source code. They both are translated into the corresponding mathematical models in two steps. The intermediate representations, a conceptual scheme plan and GVF (Generic Verification Framework), are needed for practical reasons, mainly, to allow us to deal with idiosyncrasies in a particular input notation, common in railway scheme plans, and to enable a generic verification approach relying on symbolic execution.

The two formal models derived from the scheme plan and SSI GDL are interrelated using the derived safety invariants, which leads to generation of necessary verification conjectures. Any violations of the safety invariants are reported as findings in the final report. All model transformations during steps 0-9 in figure 1 are traceable in both directions via meta-references created during the tool operation to allow the tool to create the final report describing the findings in terms of the initial inputs: for each violation it includes its main characteristics as well the relevant part of the SSI code and the schema plan fragment (several extracts of the verification reports can be found in (Iliasov et al 2018, Taylor et al 2019).

3 Industrial experience

In the last two years SafeCap has been extensively used to automatically verify the compliance of real-world SSI GDL data with safety properties. 26 different UK interlockings, developed by several different suppliers and multiple design offices, have been verified. Six of these interlockings were analysed as trial applications of SafeCap, demonstrating that the automated approach could consistently find known errors in the data (both deliberately seeded and unintentional in non-in-service data sets) in far less time than it would take to check or test manually (minutes rather than weeks). The trial applications also showed SafeCap's value in highlighting vulnerabilities in data, where safety depended on the complex interaction of different signalling principles applying to different sections of a railway layout in ways that would not be immediately apparent to a manual designer or checker charged with modifying that data. The remaining 20 were commercial applications of SafeCap in live signalling projects fulfilling industry requirements for automated verification.

Safety property violations, reported by SafeCap in these practical applications, fell into four categories:

- errors – straightforward errors that needed to be corrected;

- vulnerabilities – where the combination of circumstances under which a reported violation occurs can, through other properties, be shown impossible in practice, though it could unintentionally be made possible through modifications to seemingly unrelated data;
- intentional violations – violations of specific properties in specific locations for operational reasons, for example to allow trains to shunt backwards and forwards in a siding without signaller involvement or to allow a train to couple to the front of another train;
- false positives – reported violations where it can be shown that none exists, resulting from limitations of the safety properties as explained below.

Practical experience of SafeCap with real-world data has led to refinements of safety properties to reduce false positives by reflecting the manner in which SSI GDL data is constructed. To make efficient use of once scarce processing power, signalling interlockings do not explicitly test signalling principles in every instance they apply. Instead, they test specific instances and infer compliance in other instances through other signalling principles. Constructing a logical proof of compliance with one signalling principle can require safety properties that explicitly exclude cases that violate other signalling principles and additional safety properties to prove those other principles.

For example, where multiple set of points that are wired to always move together, the interlocking may only test that the last section of route (known as a *sub route*) locking any of the points is free. The fact that other the sub route(s) locking the points is also free is inferred, because it is locked at the same time as, and freed sequentially before, the sub route being tested. This is illustrated in figure 3.

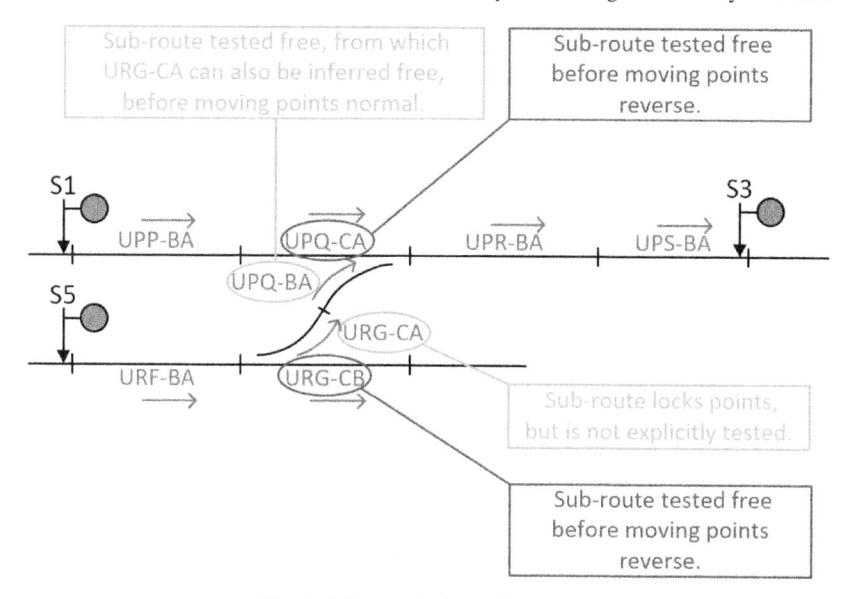

Fig. 3. Sub route locking of points

Sometimes interactions between signalling principles can be even more complex, involving interactions between points and overlaps in apparently unrelated parts of the layout. Whilst such cases can still be proven to be safe by SafeCap, they are unlikely to be obvious to anyone modifying the data and hence represent a safety vulnerability to which SafeCap can alert designers.

Even though the fully automated SafeCap verification does not require any manual proofs, is still accompanied by some manual efforts from the formal method expert and the experienced signalling engineer. In some situations, the former needs to adjust the proof tactics and properties for a specific project, the latter is involved in adjusting the properties and in the interpretation of the automated verification results. The railway expert leads the production of the final report including the categorisation of the violations found and the selection of the counter examples. Some manual efforts are often required to deal with differences in the way the SSI programs and the schemas are produced by different engineering teams and with the mismatches between the SSI program and the schema (typically, naming or naming convention mismatches) in a given project.

Our current aim is to substantially reduce these manual efforts. For example, relying on our significant experience in applying the tool we have already developed a limited set of possible proof tactics, from which we could quickly select the most suitable one when necessary.

The development or re-development of the railway network, including signalling, is naturally structured into projects. Each project constitutes a substantial geographical area (such as a large UK station), controlled by 2-4 interlockings (or virtual interlockings[1]) each of which has an SSI program located on one interlocking controller unit. This design practice restricts the complexity of individual SSI programs; however, our experience shows that it can still vary a lot depending on the way the individual SSI program is written.

Table 1. Statistics about verification in three projects

Name	Routes	Points	Signals	Safety invariants	State transitions	Time, seconds	RAM memory, peak
X	140	48	83	55	11078	37	42Gb
Y	64	29	92	55	4462	3	4Gb
Z	190	64	84	55	112560	265	67Gb

Table 1 shows the statistics about the verification conducted in the last three recently completed projects run on a professional PC with 16 cores. The Z project was one of the projects we completed with the longest verification run. We note here that the scheme elements are not a good predictor of verification complexity; instead complexity, especially its upper boundary, is better explained by the maximum cyclomatic code complexity of the verified SSI program. In practice this is often defined by cascading swinging overlaps that tend to require deeply nested subroutine calls. The state space of these SSI projects is extremely large as they have in average 1.5-2.5K Boolean variables and 100-200 Integer variables, but this is irrelevant to the type of verification we conduct.

As part of our work, we have extended the toolset by importing several proprietary schema notations used by our industrial partners and successfully experimented with importing schemas in RailM[2] and SDEF[3].

During our work on the live projects, we continue the improvements of the toolset. First, we have extended the number of the safety invariants to improve the verification coverage. Secondly, we have improved the quality of reporting by reducing the number of false positives. Substantial efforts have been dedicated to improving the quality of the safety invariants with the aim of reducing the

---

[1] A virtual interlocking is a specific interlocking application running on a shared hardware platform, often with other virtual interlockings. It differs from a traditional electronic interlocking in which there was a one-to-one mapping between interlocking applications and the hardware they ran on.

[2] Railway Markup Language. RailML.org. The RailML Standard. Data exchange v.3.1. The RailML.org Initiative. https://www.railml.org/en/introduction.html Accessed 26 August, 2021

[3] Signalling Data Exchange Format. Network Rail Signalling Innovations Group (NR SIG). SNIP – 132890 System Data Exchange Format (SDEF). Version 7.2 Design. 2014 https://www.sparkrail.org/Lists/Records/DispForm.aspx?ID=24659 Accessed 26 August, 2021

verification time and improving the diagnostics of the property violations. During this work, while dealing with a number of interlocking datasets designed by different companies and their different offices, we have developed a library of proof tactics that can improve the proof automation. We continue putting substantial efforts into improving the scalability of the tool to make sure that, even with the extended number of the safety invariants, it takes no longer than 5-6 minutes to verify the safety of any UK signalling interlocking. Finally, we have greatly improved the usability of the native SafeCap schema editor as, in our experience, the vast majority of railway schemas in the UK are available only as PDF images or CAD files from which signalling information cannot readily be gleaned.

Initial commercial applications of SafeCap have been in response to a new UK industry requirement for automated verification, which provides additional mitigation of the risk of error in safety-critical SSI GDL data. However, SafeCap has the potential to offer much greater benefit if used earlier in the design process. Doing so would not only meet the industry requirement for automated verification, but could also enable earlier identification of errors in the data production process, thereby avoiding expensive and time-consuming rework cycles. We are currently working with industry partners on determining the optimal phases for SafeCap analysis within the existing data production process.

4 Building a safety case

SafeCap is currently used in an advisory capacity, supplementing manual checking and testing processes by providing an additional level of verification and enabling earlier identification of errors. Whilst this brings significant benefits, greater value could be realised by using SafeCap as an alternative to these costly and time-consuming manual processes. However, to do this it is necessary to demonstrate that it is safe to use SafeCap in this manner.

Specifically it is necessary to demonstrate that each hazard and associated risk arising from the use of SafeCap in a signalling data preparation process is controlled to an acceptable level. The stringency of the acceptability criteria increases with the dependency placed on SafeCap within the data production process. The current use of SafeCap, as an advisory tool, places minimal dependency on it as proven manual processes remain in place. Greater dependency is placed on it if it is used as an alternative to manual checking or testing. Even greater dependency is placed on it if used as an alternative both manual checking and testing. An iterative approach is being taken to safety case development whereby the case will be progressively extended in stages, as the tool itself develops and matures, to enable greater dependency to be placed on it. The decision to use SafeCap as an alternative to specific manual checking and / or testing processes will be made be made by signalling system suppliers developing data using these processes, informed by a compelling safety case that safety risks are being suitably managed.

The safety case is being developed in accordance with the process outlined in the European Common Safety Method for Risk Evaluation and Assessment (CSM-REA) (European Parliament and Council 2015). The first step of this process is to assess whether a change – in this case the use of SafeCap in data production processes – is deemed 'significant' from a safety management perspective. The authors' assessment is that current use of SafeCap in an advisory capacity is not a 'significant' change, as SafeCap merely supplements existing processes. However, replacing established manual checks / tests by SafeCap verification is believed to be significant and hence necessitates the application of the full CSM-REA process. This is because of the novelty of the SafeCap approach, the complexity of the change and the credible worst-case scenario failure consequence that a serious error in signalling data goes undetected (a 'false negative').

Having established the need to apply CSM-REA, the next step is to define the system for which the safety case is being produced. The System Definition is required to cover the system objectives, system functions and elements, system boundary, physical and functional interfaces, system environment, existing safety measures and assumptions. The safety case, and associated System Definition, for the SafeCap tool, is being developed independently of the properties that SafeCap verifies; the correctness of the safety properties is demonstrated through their traceability to established signalling principles. This enables development of the tool and its associated safety case to be de-coupled from the safety properties.

As the SafeCap tool continues to develop and the dependency placed on it increases, the system description will need updates to reflect the latest objective and construction of SafeCap. Similarly, as safety properties are refined or expanded or new sets of safety properties are produced to cover different technologies / signalling principles so traceability to established signalling principles for specific railways can be undertaken independently of the tool.

Safety hazards (for the SafeCap tool) are identified at the system boundary, as documented in the System Description, and recorded in a SafeCap hazard log. Principle among these are the interpretation of SSI GDL and reporting of safety property violations; failings in either could lead to serious error in signalling data going undetected. Various mitigation measures are being developed to reduce the risk associated with these, and other hazards to a level assessed to be either broadly acceptable or as low as reasonably practicable (ALARP).

In the case of safety properties, the safety argument hinges not on explicit risk assessment as for the tool, but on demonstrating that the safety principles embody specific signalling principles. This approach is referred in CSM-REA as risk acceptance by application of code of practice, the codes of practice in question being the standards in which signalling principles are specified. The argument for the correctness and completeness of these principles stems from the extensive

Formal verification of railway interlocking and its safety case 295

expert review, risk assessment and real-world operational experience, over more than two centuries of railway history that has led to them in their current form. As explained earlier, attempting to improve on this through formal demonstration of the absence of hazards is almost hopeless due to the many real-world limitations.

There is, however, a need to demonstrate that these signalling principles have been correctly encapsulated in formal notation. At a basic level, this is achieved through requirements tracing: identifying the specific safety property(ies) that implement a safety property as expressed in specific clauses in railway standards. However, as explained in Section 3, implementation of one signalling principle can depend on others being upheld. In such cases, goal structured notation is used to demonstrate how the goal of demonstrating compliance with a signalling principle is upheld through lower level safety properties, as in the example in figure 4.

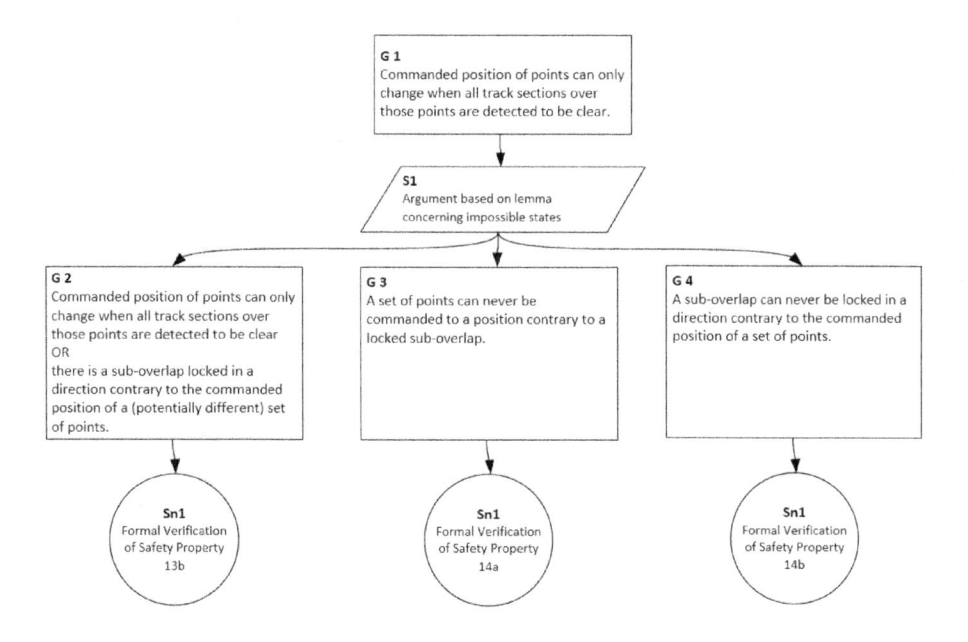

Fig. 4. Goal Structure Notation for example signalling principle

The safety argument also requires application constraints that must be adhered to for the traceability to be valid:

- trackside signalling equipment must be installed in accordance with the signalling plans provided for SafeCap verification;

- interlocking hardware inputs and outputs must be wired consistently with the functions assigned to them in data;
- points that are shown as always moving together on signalling plans must be wired to do so in the real-world;
- etc.

5 Conclusions

This paper reports on our successful and substantial deployment of formal methods in the railway industry. The SSI SafeCap tool developed for the verification of the SSI signalling meets the ambitious requirements we set when we started this work (Section 3): the tool uses the signalling and schema notations used by industry, its application does not require any knowledge of formal methods from signalling engineers, the verification is fully automated and hidden from the engineers, the tool is scalable and capable of verifying any mainline signalling within minutes. The tool has been deployed in industry: it is being applied in a number of live signalling projects now.

The research work underlying the development on this tool has been solely driven by the aim of achieving its successful industrial deployment. It has been our principal position from the outset that only with this approach we achieve real deployment of formal methods.

Our future plans focus on improving and extending the safety properties, building a safety case for the tool through which to gain its acceptance as an alternative to manual checking / testing and improving the quality of diagnostics. In the longer term we plan to use the generality and openness of SafeCap to extend the tool with verification of other signalling representations used in metro and mainline signalling in the UK and overseas.

References

Badeau F, Amelot A (2005) Using B as a High Level Programming Language in an Industrial Project: Roissy VAL, in: Proceedings of ZB 2005: Formal Specification and Development in Z and B, LNCS 3455, Springer. pp. 334–354

Cribbens A H (1987) Solid State Interlocking (SSI): an integrated electronic signalling system for mainline railways, Proc. IEE 134 (3) 148-158

Des Rivieres J, Wiegand J (2004) Eclipse: A platform for integrating development tools. IBM Systems J 43(2):371-383

European Parliament and Council (2015) European Parliament and Council Regulation (EU) 402/2013 on the common safety method for risk evaluation and assessment, Official Journal of the European Union, 30th April, 2013 as amended by Regulation (EU) 2015/1136 of 13th July, 2015

Iliasov A, Lopatkin I, Romanovsky A (2014) Practical Formal Methods in Railways – The SafeCap Approach, in: Proceedings of Reliable Software Technologies (Ada-Europe), LNCS 8454, Springer. pp. 177–192

Iliasov A, Lopatkin I, Romanovsky A (2013) The SafeCap Platform for Modelling Railway Safety and Capacity, in: Proceedings of SAFECOMP - Computer Safety, Reliability and Security. LNCS 8135, Springer. pp. 130–137

Iliasov A, Romanovsky A (2012) SafeCap domain language for reasoning about safety and capacity, in: Proceedings of Workshop on Dependable Transportation Systems at the Pacific-Rim Dependable Computing Conference (PRDC 2012). Niigata, Japan. IEEE CS, pp. 1-10

Iliasov A, Taylor D, Laibinis L et al (2018) Formal Verification of Signalling Programs with SafeCap, in: Proceedings of 37th International Conference, SAFECOMP 2018, Vasteros, Sweden, September 19-21, 2018, LNCS 11093, Springer. pp. 91–106

Iliasov A, Taylor D, Romanovsky A (2018) Automated testing of SSI data. IRSE (Institution of Railway Signal Engineers) News 241, February 2018. https://www.irse.org/Publications-Resources/IRSE-News/Archived-Issues Assessed 10 September 2021

Network Rail (2015) Company Standard NR/L2/SIG/30009/GKRT0060 'Interlocking Principles,' Issue 2, 07/03/2015

Taylor D, Iliasov A, Romanovsky A et al (2019) Driving Efficiency & Resilience to Human Error: SafeCap Automated Verification of Signalling Data, in: Proceedings of IRSE ASPECT 2019, Delft, Netherlands, October 21-24. IRSE. https://webinfo.uk/webdocssl/irse-kbase/ref-viewer.aspx?refno=740881177 Accessed 20 August 2021

Safety-critical Multi-core for Avionics

Gary Gilliland

DDC-I Inc.

Phoenix, AZ

Abstract *Creating a multi-core platform for safety-critical avionics is the next major step for most avionics manufactures. While multi-core processors are commonly used in most other markets, the avionics industry has taken years to trust multi-core technologies. Acceptance has been slow due to an avionics system's stringent safety and deterministic requirements. As a result, years of study have been invested by certification authorities and industry suppliers to identify the issues multi-core processors pose for safety-critical systems. Formalized positions of these efforts are the FAA CAST-32A Positioning Paper, and EASA's multi-core Certification Review Item (CRI). The crux of these papers (regarding software) focuses on bounding and controlling the interference patterns that exists when processor cores share resources. This paper highlights the challenges of implementing multi-core processors for avionics developers. It will present Deos SafeMCTM and show how it helps address CAST-32A objectives by utilizing unique operating system features designed for minimizing and bounding contention issues within multi-core environments. Features such as cache partitioning, memory pooling and safe scheduling enable the user to configure the memory architecture to minimize cache thrashing and schedule applications across all cores. Further, most of these Deos features are processor agnostic which allows system developers to pick more current and best suited processor technologies. Together, these capabilities enable developers to employ modern systems that orchestrate software applications such that conflicts over shared resources are minimized and the overall performance advantages of multicore processors can best be utilized.*

© DDC-I 2022.
Published by the Safety-Critical Systems Club. All Rights Reserved.

1 Introduction

Providing multi-core platforms for certifiable safety-critical avionics applications is the next major step for avionics manufacturers. These manufactures are developing systems utilizing ARM, PowerPC and IA64 architectures with anywhere from 2 to 8 cores. Deploying multi-core platforms, where all cores are active and capable of running high design assurance level (DAL) safety-critical applications, present many challenges to the developer. The guidance from the certifying authorities (FAA, EASA, TC, etc.[1]) is not prescriptive and therefore requires the avionics manufacture to create a case for certification based on the guidance.

Further complicating the use of multi-core is the lack of specifically designed common off the shelf (COTS) multi-core processors for safety-critical avionics. Processor manufactures such as NXP, Texas Instruments, ARM, Xilinx, etc. tend to focus on markets where they can sell millions of devices, typical volumes in aerospace programs are in the hundreds. Fortunately, the automotive market which has higher volumes and interest in safety-critical systems helps drive processor architectures toward what is needed in avionics. Regardless of the architecture it is the responsibility of the system integrator to make these multi-core processors work in a safety-critical environment.

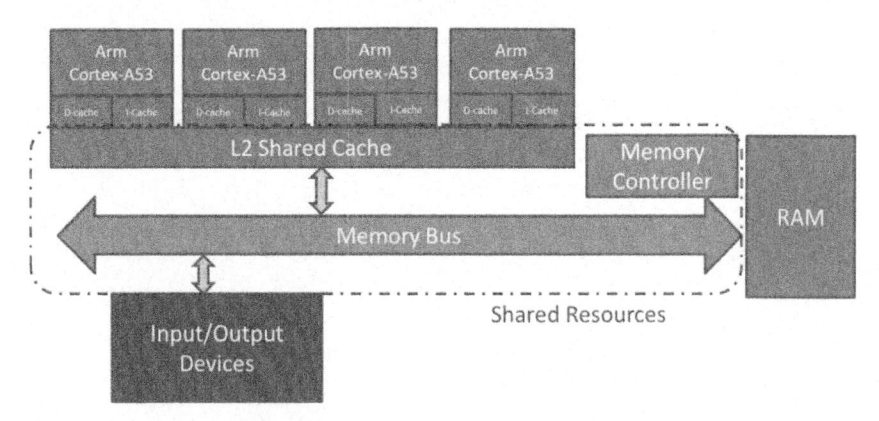

Fig. 1. Generic Multi-core Architecture

Multi-core processors (MCPs) have become increasingly prevalent. They offer increased computing power without significant increases in size, weight and power, which makes them appealing to the developers of embedded applications. However, figure 1 shows a representative MCP architecture and the area of

---

[1] Different parts of the world have different certification authorities, the United States has the Federal Aviation Authority (FAA), Canada's authority is called Transport Canada and the countries in the European Union use the European Union Aviation Safety Agency (EASA).

shared resources that present two key challenges to the developers of certifiable, safety-critical applications:

- MCP's cores share critical resources, such as L2 cache and a memory subsystem (i.e., a memory bus and RAM). Since software executing on different cores is contending for usage of these shared resources, an MCP greatly increases the potential for interference patterns whereby software on one core can impact the execution time of software on another core. Consequently, the impacted software may miss deadlines, resulting in unsafe failure conditions.
- MCP's are designed to optimize average-case execution time, often at the expense of worst-case execution times. Since developers of certifiable, safety-critical software must design for worst-case behaviour[2], the delta between worst-case and average-case execution times is typically inflated.

Using an MCP in a certifiable, safety-critical application requires one to:

- Bound and control interference patterns on shared resources. Without this capability, safe operation cannot be guaranteed and certification is impossible.
- Productively use budgeted but unused execution time. Without this capability, CPU utilization will be low, thereby wasting much of an MCP's computing capacity.

Developers of certifiable, safety-critical applications must solve these two challenges to ensure that the systems they deliver will safely perform their intended functions. Additionally, developers must explain to the certification authorities exactly how they use MCPs in their systems and how they ensure safe operation. Organizations such as EASA and the FAA[1] are becoming increasingly aware of the challenges posed by MCPs and MCP-based systems will be subjected to intense scrutiny during the certification process. To that end, the Certification Authority Software Team (CAST) which is made up of certification authorities from North and South America, Europe, and Asia has published CAST-32A, a position paper on MCP-based systems.

2 CAST-32A Guidance

CAST-32A (CAST-32A, 2016) identifies some specific topics that need to be address by your design. While the operating system can enable the developers to meet these objectives, it is up to the system integrator to properly configure the operating system and applications to meet these requirements. According to CAST-32A planning, testing, verification and reporting are important parts of the process. These processes are all focused on the interference channels inherent in MCP systems, and how they are going to be mitigated.

---

[2] Safety-critical software must be able to meet its deadlines, even under worst-case conditions. If it can't, it may result in an unsafe failure condition.

The Deos Operating System, since its inception in 1995, has been designed, implemented and verified to provide both *Robust Resource Partitioning*[3] and *Robust Time Partitioning*[2] as defined by CAST-32A. Deos multi-core provides *Robust Resource Partitioning* and *Robust Time Partitioning* by giving the target system developer *interference channel*[2] solutions that range from elimination of the *interference channel* to a definitive bound on the *interference channel*. Therefore, addressing the concerns of CAST-32A is achieved by the following steps:

1. Define the *interference channels* in the target system
2. Specify how those *interference channels* are eliminated or bound using Deos provided features
3. Implement *interference channel* eliminations and bounds using Deos provided features
4. Verify the correctness of the implementation
5. Include appropriate *Safety Nets*[2] as directed by the target system's safety analysis
6. Document what has been accomplished

It is also important to note that there are some key recommendations mentioned in the CAST-32A position paper.

- MCP platforms on which software applications or threads can be dynamically re-allocated to a different core (or different cores) by the operating system. This implies that threads of execution must be configured to run on a specific core. (i.e. Threads should have core affinity).
- The use of CPU multithreading or hyper-threading. Most modern CPUs have the ability to execute multiple threads of execution on a single physical core. The CPU hardware only has a single set of execution resources for each core; therefore these threads are sharing the hardware resources. Since this is all happening in the internal logic of the CPU there is not any good way of controlling or bounding the interference that could occur. As a result, it is recommended that the multithreading capability should not be used in a DAL-A system.

---

[3] This term is defined in CAST-32A. Refer to that document for a definition.

3 Multi-core Challenges

3.1 Resource Contention and Interference Patterns

Figure 2 shows a simple dual-core processor configuration, each core with its own CPU and L1 cache, and both cores sharing an L2 cache and a memory subsystem. The primary areas of contention are shown in yellow:

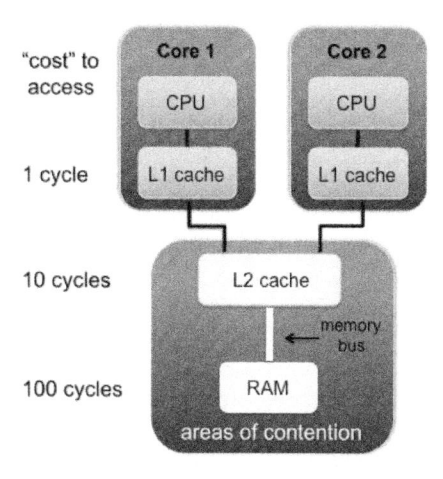

Fig. 2. Simple Dual Core Configuration

The values shown at the left of the figure represent the "cost" that each CPU incurs when accessing a given resource. For example, say it costs one cycle for the CPU to access its local L1 cache. If the L1 access misses and the CPU has to access the L2 cache, it costs 10 cycles. If the L2 access misses and the CPU has to access RAM, it costs 100 cycles. Further, if either cache is "dirty"[4] or a "write-back"[5] is needed, the cost is even greater. The fact that a memory access may cause both a cache line read for the memory access and an additional cache line

---

[4] Cache is a high-speed data area which stores the latest text or data area of an application. Cache is used so the CPU doesn't have to retrieve the data from RAM which speeds of processing. Upon a context switch if the required data is not in the cache, the cache is considered dirty and must be flush and filled with the correct RAM area.

[5] Cache has 2 typical modes of operation, write-through and write-back. In write-through the write is done synchronously to both the cache and the RAM so what is in the cache is in RAM. In write-back mode writing is only done to cache and not transferred to RAM until it is about to be ejected from cache. When it is determined that the cache is dirty the cache must be written to RAM and then flushed and refilled with the required data. This typically takes longer that just a flush and refill as required for write-through.

write due to the cache conflict is a major source of execution time variability. Note that these cost numbers aren't intended to be exact and will vary between processors and board designs, but the relative orders of magnitude are typical. The point is that the further out the CPU has to reach to access data, the more time the data transfer takes. In addition to cache contention, software on different cores can contend for access to the underlying memory subsystem (i.e., the memory bus and RAM), thereby creating undesirable interference patterns. For example, if the memory subsystem is servicing a request from software on core 0, then accesses by software on core 1 will be forced to wait.

In addition to cache contention, software on different cores can contend for access to the underlying memory subsystem (i.e., the memory bus and RAM), thereby creating undesirable interference patterns. For example, if the memory subsystem is servicing a request from software on core 0, then accesses by software on core 1 will be forced to wait.

In summary, at the L2 and memory subsystem level, contention and interference patterns arise when software on multiple cores compete for those shared resources.[6] When such interference occurs, one software module can cause another module to execute more slowly than normal, which poses serious problems in a certified, safety-critical application.

3.2 Software Design Induced Contention

When designing a real-time operating system (RTOS) for a multi-core platform or the software applications to run on this RTOS one must take care that the implementation does not cause any contention between cores. From an RTOS perspective this is related to how the RTOS manages the scheduling, the resources it needs to protect regions of code (semaphores, mutexes, etc.) as well as how it allocates resources to applications.

Application software typically also has needs to protect regions of code or coordinate execution of threads with the use of semaphores and mutexes. Application software can also communicate with other applications in the MCP system. Since these applications can be executing on different cores care must be taken that the implementation does not cause any undue contention between cores.
It is important that the kernel resources used to perform these actions are available on a per-core basis such that it does not block another core in order to create and use these resources.

---

[6] Even on a single core, multi-process partitions compete for that core's L1 cache, thereby creating contention and interference patterns at the L1 level as well.

3.3 Time Budgeting and Slack

In safety-critical systems, a process must be allocated a time budget that is adequate for its worst-case execution time (WCET). Even if the L1 and L2 caches are dirty (from the process' perspective) and there is contention for memory, the process must have sufficient time budget to complete its intended function. Otherwise, unsafe conditions may arise quickly, up to and including loss of equipment and human life.

However, worst-case execution scenarios rarely occur and a process usually experiences something close to its average case execution time (ACET). As a result, if a process' time budget is set for its WCET, the typical result will be a significant amount of budgeted, but unused, time. For example, consider a process that is given a 500 μsec time budget (WCET). If its ACET is 300 μsec, then on average, it will generate 200 μsec of budgeted but unused time per execution period. In Deos, this "extra" time is called "slack".

Consequently, the amount of slack time generated in a single-core, safety-critical application can be substantial. Developers often see 70% or less of a CPU's cycles utilized, even though the CPU is budgeted at over 90%. In an MCP-based application, this problem is exacerbated since WCETs are typically inflated due to increased resource contention and interference patterns amongst the cores. This concept of slack utilization is unique to Deos and is a patented capability.

In summary, in safety-critical applications, WCETs are often significantly greater than ACETs due to resource contention and interference patterns. Unless this budgeted but unused "slack" time can be harvested and put to good use, CPU utilization will be seriously degraded, thereby wasting a limited resource.

4 Resource Contention Approaches

There are many approaches that have been employed, which attempt to solve the multi-core challenges we outlined earlier. In this section, we discuss these approaches.

4.1 Cache Policies

Traditionally, three caching policies have been used to reduce contention and minimize the delta between a thread's WCET and its ACET.

- **Disable L2 (shared) cache** - The most rudimentary policy involves disabling the cache, which greatly reduces the difference between a thread's WCET and

its ACET. Unfortunately, this policy cripples the CPU, as every cache access becomes a miss and requires an external RAM access.
- **Cache Locking** - this approach involves locking critical code or data into given cache lines or to particular cores. This policy ensures high cache hit rates for a relatively small amount of application code or data, but the rest of the software must compete for the remaining non-locked cache, with increased contention. Essentially, cache locking picks winners and losers, and most are losers. This methodology is very processor dependent and application specific and therefore not portable. As a matter of fact, some of the latest processor architectures don't give the user the ability to lock cache.
- **Cache Flushing** - this approach involves flushing the cache prior to entering a new process. That way, each process starts with an empty cache, then begins to fill the cache as it executes so that it benefits from increasingly higher cache hit rates. The problem with cache flushing is that processor manufacturers optimize caches for desktop/server/datacom applications. In these non-safety-critical systems, the minimum time quantum is usually 10 msec, so once an application gets control of the CPU, it runs for at least 10 msec and usually much longer before the Operating System (OS) preempts it. This gives the software time to prime the cache and achieve a very high cache hit rates (>99%) thereafter. In real-time safety-critical systems, by contrast, dozens or hundreds of processes typically execute within a 10 msec window. As a result, these processes don't have the time to prime the cache in their favor, so they experience much lower cache hit rates and rely more on the memory subsystem. Also, as caches become larger and larger, the act of flushing the cache becomes more and more expensive (e.g., many 10's of micro-seconds). This cost imposes a significant "pause" before each process can begin execution.

While cache flushing is better than simply disabling the cache, it is not very effective and incurs a significant performance penalty for real-time safety-critical applications.

In summary, the RTOS must provide an efficient method of managing the shared cache. The solutions provided in the marketplace vary widely. Some recommend disabling the cache, others suggest using hardware capability to partitioning the cache on a per-core basis or partitioning the cache using user level cache control. Deos provides a way to partition the cache on a per-application basis that will be discussed in the next section.

5 DDC-I Solution for Multi-core

SafeMC™ technology for Deos[7], a safety-critical real-time operating system for multi-core processors, provides the resource and scheduling mechanisms that enable developers to bound and control the interference patterns which occur whenever processor cores share resources (e.g., cache, memory, or I/O). This capability enables Deos developers to have full utilization of all cores for safety-critical operation, as opposed to artificially forcing the user to designate particular cores to particular Design Assurance Levels (DALs).

The Deos multi-core technology provides a solution to the resource contention issues by bounding and controlling interference patterns with a few key technologies.

5.1 Memory pooling and cache partitioning

Memory pooling and cache partitioning is a configurable cache partitioning capability that enables developers to isolate cache at the application or partition level (not just at the core level) via software configuration files (not dependent on hardware features). This patented Deos feature substantially reduces one of the major sources of resource contention, namely access to cache and system memory.

- Deos memory pooling and cache partitioning is a capability that enables the user to configure separate areas of cache for each application as shown in Figure 3. Since no other application will use this area of cache it negates the need to flush cache between partition context switches. Cache flushes become very expensive time-wise when using devices with large L2/L3 caches. This partitioning reduces the total amount of L2 cache available to each core; however, overall contention is reduced, as multiple cores no longer compete for the same resource which improves performance.
- Deos memory pooling feature allows developers to define where applications run in physical memory. Depending on the processors and architecture considered, memory pooling could be setup to use on-chip high speed RAM and run specific critical applications from that memory.
- If the application developer does not share a memory pool with a DAL A process or a process on another core, there will not be any contention.

---

[7] Deos is a time, space and resource partitioned RTOS designed for certifiable, safety-critical applications. It has been certified to DO-178C Level A design assurance and is used in many avionics functions on many aircraft.

- Memory pooling and cache partitioning, settings are set through the platform XML configuration files (not discrete hardware settings), which greatly simplifies system development, design, testing, tuning, and future reconfigurations.

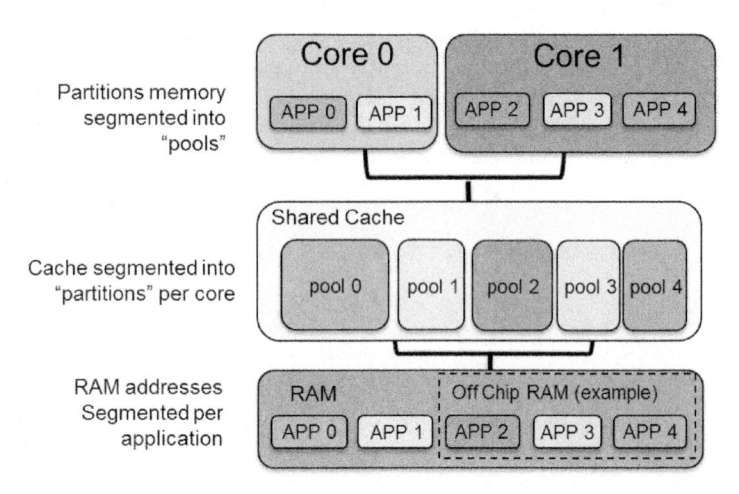

Fig. 3. Memory Pools and Cache Partitioning

It is important to note that the use of cache partitioning and memory pooling capability does not change the way the hardware cache works. It works as intended just on a smaller cache size, so even if you want to use some of the hardware cache locking mechanisms or turning cache into high-speed RAM, Deos could work in conjunction with those hardware configurations. In fact, memory pooling could be setup to use that cache as high-speed RAM to run specific critical applications from that memory. Memory pooling and cache partitioning can also be used on processors that have multiple memory controllers. The system can be configured such that all memory attached to one memory controller is assigned to applications on one core and the memory attached to the other memory controller is assigned to applications on difference core. The cache can be divided and allocated between the two cores as well. Deos cache partitioning is the first line of defense for bounding hardware contention due to memory and cache accesses from multiple cores. This software solution is portable to all supported hardware architectures.

5.2 Safe Scheduling

Deos's safe scheduling includes patented technology for the bounding and control of interference patterns created by the shared resources within multi-core processors. Deos uses a two-level scheduling model which employs a time window scheduler that is synchronized across all cores as well as a scheduler per core. The first level scheduler manages time in windows which are aligned across all cores. Within a time window a second level scheduler (i.e., ARINC-653, Deos RMS POSIX, Ada[8]) is assigned to each core and then applications are assigned to one, or more, of the schedulers. The goal of this architecture is to allow the developer to orchestrate the scheduling of applications running on different cores and different windows such that they have limited interference with each other. Safe Scheduling offers a configurable approach of a single RTOS instance managing all the cores and provides user control over the co-scheduling of tasks among the cores. As shown in Figure 4, this scheduling environment allows high-DAL applications to run simultaneously across all cores by enabling the system integrator[9] to configure the system such that resource contentions are minimized (and bounded) between processor cores.

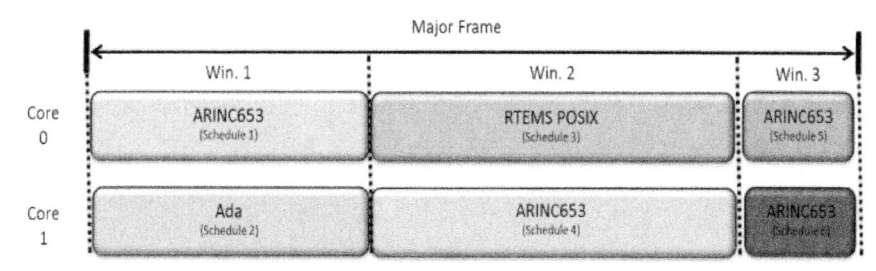

Fig. 4. Safe Scheduling

The system engineering process for configuring a multi-core system involves an analysis of all the applications that are required to run on the system. The integration of these applications involves determining the resource requirements for

---

[8] The second level schedulers can be any one of the provided schedulers. The name of the scheduler implies the API's that are used for application as well as the scheduling of those task.
- ARINC653 is a industry standard API and scheduling model
- RMS is the Deos native API and rate monotonic scheduling (RMS) model
- POSIX is an industry standard API which utilizes the Real-Time Executive for Multiprocessor Systems (RTEMS) scheduling model
- Ada is an industry standard programing language that utilizes the Deos native RMS scheduling model.

[9] In an avionics system the system integrator is the one responsible for getting all the software for a particular system to work together and form the desired solution.

each application as well as how the application uses these resources. Once this is determined the integrator can schedule the applications. Some general guidance would be that:

1. Applications that can benefit from parallel processing should be scheduled on multiple cores in the same time window.
2. Application that share interfaces will typically cause a lot of interference therefore they should be separated in time. (i.e. separate time windows)
3. Applications that create little or no interference can be scheduled in time windows where they best fit.

5.3 Safety Nets

A safety net is required according to CAST-32A to contain unintended functionality. Deos implements a method of contention reduction that uses the resources available on the hardware platform designed to monitor hardware behaviour.

Many of today's MCP processors have a performance monitor built into the chip that allows the application developer to setup and monitor activity that occurs on the device. Deos supports the ability to interface with the performance monitor and based on the behavioural limits that the system developer defines, disable the scheduling of a misbehaving partition. This capability is up to the system developer to define since only they know the parameters of the system needed to make these decisions.

6 Summary

To summarize, MCPs have become ubiquitous. However, they present significant challenges for the developers of certifiable, safety-critical applications. If one wishes to use an MCP in such applications, bounding and controlling interference patterns on shared resources, and effectively managing CPU utilization are essential. Without the first capability, certification of safety-critical software is impossible. Without the second, much of an MCP's increased computing power is wasted. The operating system selection and use of its capabilities to address the CAST-32A objectives can have a large impact not only on the system performance but can also greatly influence the degree of engineering effort it takes to certify a safety critical system on the processors one can select from.

Deos' unique, SafeMC technology (e.g., safe scheduling, cache partitioning, etc.), provides a means of solving these challenges, thereby allowing developers to fully exploit the capabilities of MCPs in certifiable, safety-critical applications.

References

ARINC Specification 653P1-3 – Avionics Application Software Standard Interface Part 1 – Required Services; Published: November 15, 2010

CAST-32A Certification Authorities Software Team (CAST) Position Paper CAST-32A Multi-core Processors, November 2016, https://www.faa.gov/aircraft/air_cert/design_approvals/air_software/cast/media/cast-32A.pdf

Deos: A time and space partitioned DO-178 Level A certifiable RTOS, 2017. http://www.ddci.com/products/deos

DO-178C Software Considerations in Airborne Systems and Equipment Certification, 2011

Real-Time Executive for Multiprocessor Systems (RTEMS); http://www.oarcorp.com/rtems

Rate Monotonic Scheduling (RMS); https://en.wikipedia.org/wiki/Rate-monotonic_scheduling

SCORE Ada; https://www.ddci.com/products_score/

At the interface of engineering safety and cyber security

Reuben McDonald

HS2

Abstract *A modern railway is a highly electrotechnical system with connectivity and networks inherent in its design. For High Speed 2 this means a design which is based on a confluence of difference wired and wireless networks. These networks support the delivery of safety critical, safety related, operation critical and wider business functions. This talk explains how HS2 is assessing its safety and cyber risks in an efficient manner, leveraging products and assessments from the application of the Common Safety Method on Risk Assessment to support cyber assessments based around IEC 62443 risk methodology and associated controls*

© HS2 2022.
Published by the Safety-Critical Systems Club. All Rights Reserved.

What do Byzantine Generals and Airbus Airliners Have in Common?

Dewi Daniels

Software Safety Ltd

Abstract *On 14 June 2020, all three primary flight control computers on an Airbus A330 shut down while it was landing at Taipei, Taiwan. The aircraft came to a stop only ten metres from the end of the runway. The cause was a problem well-known in computer science called the Byzantine Generals Problem, which was first described by Leslie Lamport in 1982. He presented a solution to the Byzantine Generals Problem, along with a mathematical proof of the correctness of that solution. This paper describes what happened on 14 June 2020 and how Leslie Lamport's solution would have avoided the incident.*

1 Introduction

China Airlines (CI) flight 202 is a scheduled passenger flight from Shanghai to Taipei. On 14 June 2020, it was operated by an Airbus A330 aircraft, with a total of ninety-eight passengers and crew on board. When the aircraft touched down at Taipei, all three Flight Control Primary Computers (FCPCs) shut down. This meant that the ground spoilers, thrust reversers and autobraking function all stopped working. The pilots eventually stopped the aircraft using manual braking and the aircraft came to a halt only ten metres from the end of the runway.

Runway excursions can be fatal. For example, 187 people were killed when an Airbus A320 overran the runway at São Paolo on 17 July 2007 (CENIPA 2009).

© Dewi Daniels 2022. All Rights Reserved.
Published by the Safety-Critical Systems Club.

Fig. 1. The aircraft stopped at the end of the runway (reproduced from the TTSB final report)

1.1 Acronyms and Definitions

ACM	Association for Computing Machinery
CENIPA	Aeronautical Accident Investigation and Prevention Center
CI	China Airlines
COM	Command
CVR	Cockpit Voice Recorder
ECAM	Electronic Centralized Aircraft Monitor
FBW	Fly By Wire
FCPC	Flight Control Primary Computer
FCS	Flight Control System
FCSC	Flight Control Secondary Computer
FDR	Flight Data Recorder
MON	Monitor
NASA	National Aeronautics and Space Administration
OIT	Operators Information Transmission
OM	Oral Message
PF	Pilot Flying
PM	Pilot Monitoring
PRIM	Primary
SIFT	Software Implemented Fault-Tolerance
SRI	Stanford Research Institute
TTSB	Taiwan Transportation Safety Board
WOW	Weight on Wheels

2 The Incident

The timeline of the incident is shown in Table 1 below:

Table 1. Timeline of the Incident

Local Time	Seconds Since Touchdown	Event
17:46:54	0	Aircraft touched down; ground spoilers started to deploy
17:46:57	+3	Autobrake system fault recorded on Flight Data Recorder (FDR); Electronic Centralized Aircraft Monitor (ECAM) message inhibited
17:46:58	+4	PRIM1/PRIM2/PRIM3 faults recorded on FDR (ECAM message was inhibited); spoilers retracted
17:46:58	+4	Pilot Monitoring (PM) called out "reverse"
17:46:59	+5	PF asked twice if "autobrake is on," PM answered "autobrake is not on"
17:47:04	+10	PF called out "manual brake" and applied full brake pedal
17:47:07	+13	PM called out "reverse no green"
17:47:08	+14	PF requested "quickly help me brake, help me brake"; both pilots applied full pressure on the brake pedals
17:47:22	+28	Aircraft ground speed dropped below 80 knots; master caution annunciated
17:47:36	+42	Aircraft comes to a full stop 10 m before the end of the runway

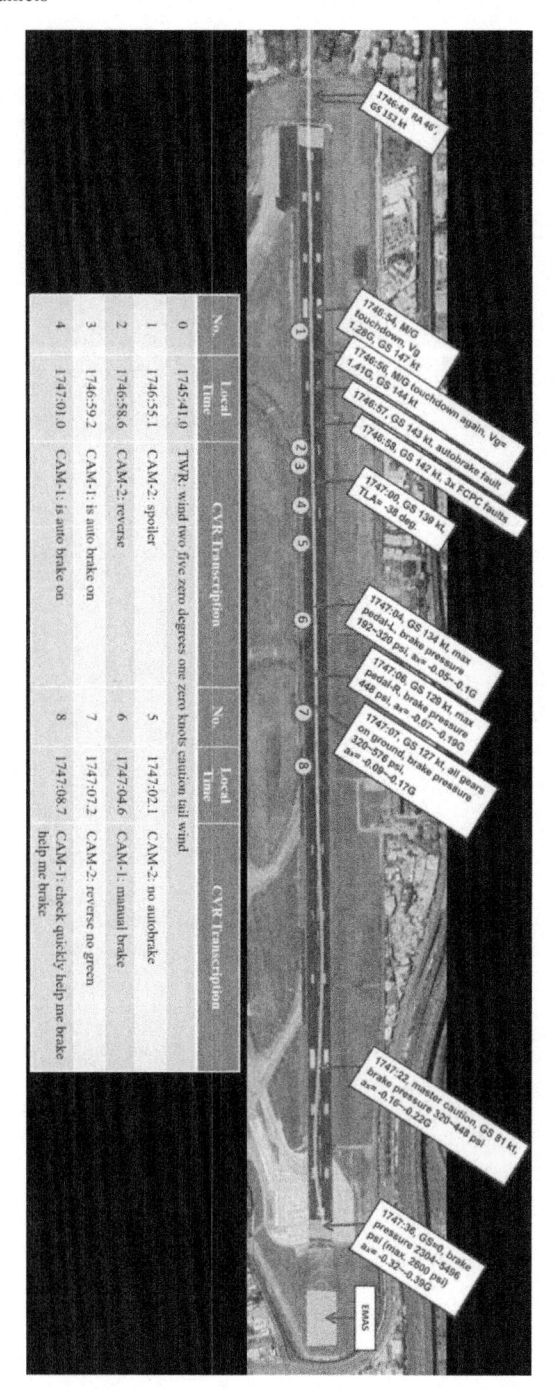

Fig. 2. Aircraft ground track, key events and CVR transcript (from the TTSB final report)

3 Cause of the Incident

The Airbus A330/A340 Flight Control System (FCS) architecture uses three Flight Control Primary Computers (FCPCs) and two Flight Control Secondary Computers (FCSCs).

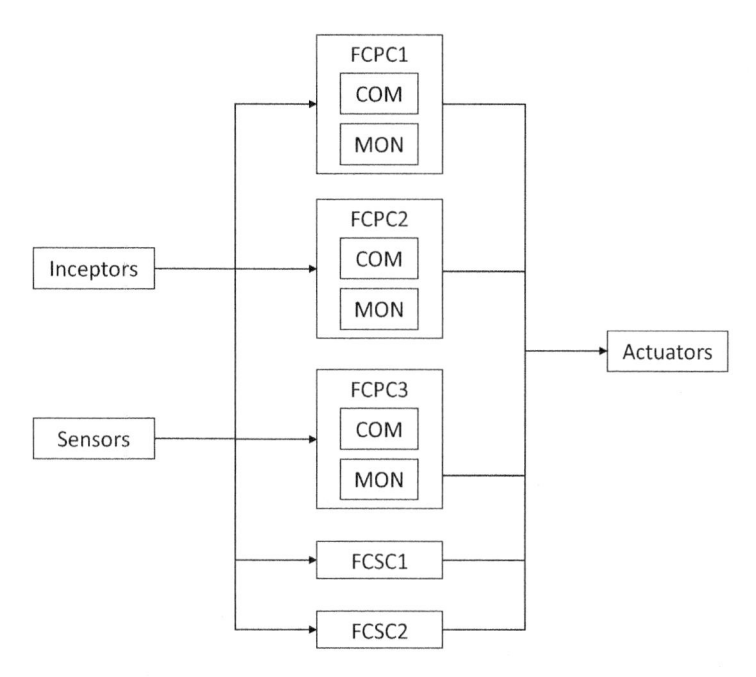

Fig. 3. Airbus A330/A340 Flight Control System

Each of the FCPCs contains a Command (COM) lane and a Monitor (MON) lane. Both the COM and MON calculate the outputs independently. Should the COM and the MON disagree, then that FCPC is shut down and control is transferred to the next FCPC. Should all the FCPCs fail, then control is transferred to the two FCSCs.

The high degree of redundancy protects against random hardware failures. The use of diverse hardware and software in the COM and MON, and the FCPCs and the FCSCs, is intended to protect against hardware or software design errors.

The Airbus A330 has three modes of operation:

1. Normal law
2. Alternate law
3. Direct law

The FCPCs implement normal law, alternate law, and direct law. The only difference between these laws that is relevant to this incident is that direct law does not provide ground spoilers, reverse thrust or autobraking; it only provides manual braking. The FCSCs only implement direct law.

In this incident, there were three factors that led to the COM/MON pairs seeing different inputs and therefore calculating different outputs, leading to three perfectly healthy FCPCs being shut down.

1. The PF happened to be moving the rudder pedals just as the aircraft touched down.
2. The aircraft bounced slightly, so that that Weight on Wheels (WOW) oscillated between true and false for about 0.75 seconds.
3. The clocks had drifted apart, so that the COM/MON pairs were reading their inputs at different times.

This caused all three COM/MON pairs in the three FCPCs to disagree as to the correct rudder output, causing all three FCPCs to be shut down in succession.

The TTSB final report concluded:

"The root cause was determined to be an undue triggering of the rudder order COM/MON monitoring concomitantly in the 3 FCPC. At the time of the aircraft lateral control flight law switching to lateral ground law at touch down, the combination of a high COM/MON channels asynchronism and the pilot pedal inputs resulted in the rudder order difference between the two channels to exceed the monitoring threshold."

This is an instance of a well-known problem in computer science called the Byzantine Generals Problem, which was first described in (Lamport et al 1982).

4 The Byzantine Generals Problem

In the late 1970s, Leslie Lamport from SRI International was working on the Software Implemented Fault-Tolerance (SIFT) project for NASA (Wensley et al 1978). SIFT was an experimental computer system intended for safety-critical avionic applications such as fly-by-wire (FBW). Leslie Lamport realised that the voting logic for a redundant computer system could fail to agree on a majority verdict even if there were no faults present. He formulated this problem as the Byzantine Generals Problem, which was described in (Lamport et al 1982).

The Byzantine Generals Problem, which was discussed in (Menon & Rainer 2022), is as follows. The Byzantine army is camped around an enemy city. The commander wishes to send orders to his lieutenants. If they all attack at the same time, they will succeed, and the enemy city will fall. If they do not all attack at the same time, they will fail, and the attacking force will be defeated. One or more of the generals is a traitor, who will try and mislead the loyal generals and cause the attack to fail.

Suppose there is one commander and two lieutenants. The commander could be a traitor, so he could command one lieutenant to attack and the other to retreat. This would cause the lieutenant who attacks on his own to be defeated.

One would suppose that the commander's treason would be revealed if the two lieutenants were to compare their orders, as in figure 3 below, in which a filled ellipse indicates the traitor:

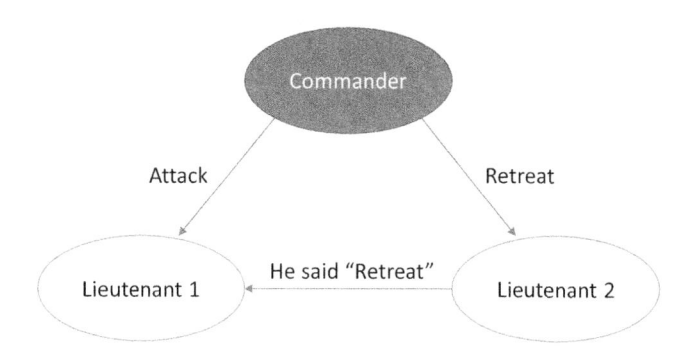

Fig. 4. The commander is a traitor

However, suppose Lieutenant 2 is the traitor. The loyal commander could have ordered both lieutenants to attack, but the traitor could claim he was ordered to retreat. See figure 4 below:

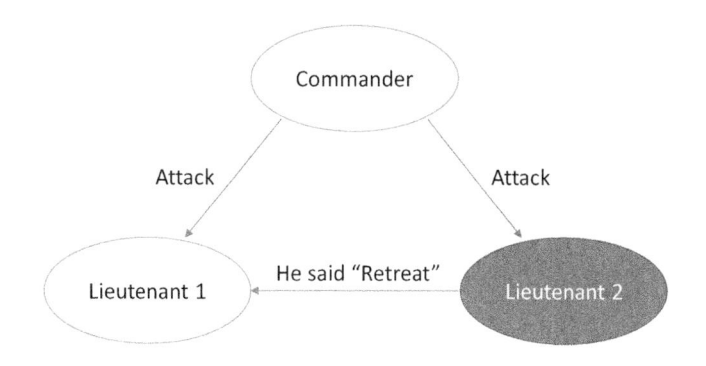

Fig. 5. Lieutenant 2 is a traitor

From Lieutenant 1's point of view, these two scenarios are indistinguishable. If the commander tells him to attack, but Lieutenant 2 tells him he was ordered to retreat, he knows that one of them is a traitor, but he does not know which one is the traitor and therefore which one to believe.

There is no solution to the Byzantine Generals Problem when there are only three generals. (Lamport et al 1982) showed that at least $3m + 1$ generals are needed to cope with m traitors. It follows that at least 4 generals are needed to cope with a single traitor.

Leslie Lamport presented a solution to the Byzantine Generals Problem, together with a mathematical proof of the correctness of that solution. He inductively defined the Oral Message algorithms OM(m), for all nonnegative integers m, which a commander sends an order to n − 1 lieutenants. He showed that OM(m) solves the Byzantine Generals Problem for 3m + 1 or more generals in the presence of at most m traitors.

Algorithm OM(0).

(1) The commander sends his value to every lieutenant.
(2) Each lieutenant uses the value he receives from the commander or uses the value RETREAT if he receives no value.

Algorithm OM(m), $m > 0$.

(1) The commander sends his value to every lieutenant.
(2) For each i, let v_i be the value Lieutenant i receives from the commander, or else be RETREAT if he receives no value. Lieutenant i acts as the commander in Algorithm OM($m - 1$) to send the value v_i to each of the $n - 2$ other lieutenants.
(3) For each i, and each $j \neq i$, let v_j be the value Lieutenant i received from Lieutenant j in step (2) (using Algorithm($m - 1$)), or else RETREAT if he received no such value. Lieutenant i uses the value *majority*(v_1, ..., v_{n-1}).

Consider the following scenario where the commander is the traitor. This time, the loyal lieutenants can figure out that the traitorous commander ordered most of the lieutenants to attack, so that is what they all agree to do.

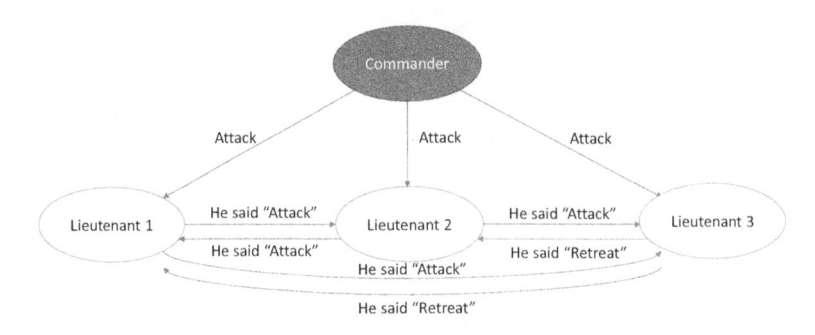

Fig. 6. Commander is a traitor (Byzantine Generals solution)

Consider the following scenario where one of the lieutenants is the traitor. This time, the loyal lieutenants can figure out that the traitorous lieutenant is lying. The loyal lieutenants both agree to attack, as ordered by the commander.

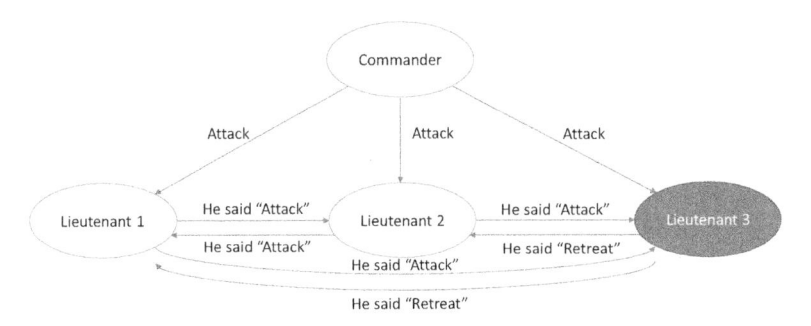

Fig. 7. Lieutenant 3 is a traitor (Byzantine Generals solution)

What is the relevance of the Byzantine Generals Problem to avionic systems? The use of majority voting, as in the Airbus COM/MON architecture, assumes that all correctly functioning processors will produce the same output. This is only true if they all use the same input. A faulty sensor may give different values to different processors. Even if the sensor is not faulty, different processors could use different values if they read the sensor at different times.

Implementing Leslie Lamport's Byzantine Generals solution ensures that all processors use the same input. A faulty sensor could still produce an incorrect value, so there still need to be sufficient redundant sensors to provide the required availability.

5 Byzantine Clock Synchronisation

It is noted that the incident occurred, in part, because the COM/MON pairs read their inputs at different times. Implementing the Byzantine Generals solution would have ensured that all the COM/MON pairs used the same input values even if they had read the inputs at different times. However, a later paper by Leslie Lamport (Lamport & Melliar-Smith 1984) described three algorithms to keep the clocks in a distributed system synchronised with one another. Two of these algorithms were derived from the Byzantine Generals solution.

6 Safety Actions Taken by Airbus

On 28 July 2020, Airbus published an Operators Information Transmission (OIT) to inform operators of the incident, which is reproduced in Appendix 4 of (TTSB 2020).

Re-architecting the A330/A340 FCS to implement the Byzantine Generals solution would have been expensive since it would have required all the COM/MON pairs in all three FCPCs to communicate between each other to exchange their input values.

Instead, (TTSB 2020) states that Airbus is implementing a software update, which will consist of several system improvements:

- *Decrease of the COM/MON asynchronism level for the flight/ground information treatment*
- *Improvement of the COM/MON rudder order monitoring robustness in case of ground to flight and flight to ground transitions*
    - *Higher unitary monitoring robustness during such transitions*
    - *Avoid cascading/"domino's" effect that leads to several PRIM fault*

Airbus claims that no other such incident has been reported since the Airbus A330/A340 family entered service. Airbus also claims their investigations have not found any other issues that could result in a similar incident. They point out that the A330/A340 fleet fitted with electrical rudder has accumulated 8.7 million flight cycles and 44.3 million flying hours.

It is noted that all Airbus designs share a very similar FCS architecture.

7 Conclusion

The Byzantine Generals Problem is a well-known problem in computer science, with a well-known solution. The Boeing 777 was designed to avoid Byzantine failures (Driscoll et al 2003, Rushby 2002 and Yeh 1998). It is surprising that Airbus did not implement the Byzantine Generals solution, which would have avoided this incident.

References

CENIPA (2009), *Final Report A – N° 67/CENIPA/2009, Aircraft PR-MBK, 17 July 2007*, http://sistema.cenipa.aer.mil.br/cenipa/paginas/relatorios/rf/en/3054ing_2007.pdf, accessed 17 December 2021.

Kevin Driscoll, Brendan Hall, Håkan Sivencrona and Phil Zumsteg (2003), *Byzantine Fault Tolerance, from Theory to Reality,* Proceedings of the International Conference on Computer Safety, Reliability, and Security (SAFECOMP 2003), Pages 235–248.

Leslie Lamport, Robert Shostak and Marshall Pease (1982), *The Byzantine Generals Problem*, ACM Transactions on Programming Languages and Systems, Vol. 4, No. 3, July 1982, Pages 382–401, https://lamport.azurewebsites.net/pubs/byz.pdf, accessed 17 December 2021.

Leslie Lamport and P. M. Melliar-Smith (1984), *Byzantine Clock Synchronization*, Proceedings of the Third Annual ACM Symposium on Principles of Distributed Computing, August 1984, Pages 68-74, https://lamport.azurewebsites.net/pubs/clocks2.pdf, accessed 17 December 2021.

Catherine Menon and Austen Rainer (2022), *Stories and narratives in safety engineering*, Proceedings of the 30th Safety-Critical Systems Symposium (SSS'22), February 2022.

John Rushby (2002), *A Comparison of Bus Architectures for Safety-Critical Embedded Systems*, CSL Technical Report, SRI International, June 2002, http://www.csl.sri.com/users/rushby/papers/buscompare.pdf, accessed 7 January 2022.

TTSB (2020), *Final Report, China Airlines Flight CI202*, Report Number TTSB-AOR-21-09-001, https://www.ttsb.gov.tw/media/4936/ci-202-final-report_english.pdf, accessed 17 December 2021.

John H. Wensley, Leslie Lamport, Jack Goldberg, Milton W. Green, Karl N. Levitt, P. M. Melliar-Smith, Robert E. Shostak and Charles B. Weinstock (1978), *SIFT: Design and Analysis of a Fault-Tolerant Computer for Aircraft Control*, Proceedings of the IEEE, Vol. 66, Issue 10, October 1978, Pages 1240–1255, https://lamport.azurewebsites.net/pubs/sift.pdf, accessed 7 January 2022.

Y.C. Yeh (1998), *Design considerations in Boeing 777 fly-by-wire computers*, Proceedings of the 3rd IEEE International High-Assurance Systems Engineering Symposium, 13–14 November 1998.

AUTHOR INDEX

Victor Bolbot 121
Evangelos Boulougouris 121
Graham Braithwaite 201
Carmen Cârlan 51
James Catmur 123
Dewi Daniels 227, 315
Mike Drennan 263
Paul Malcolm Darnell 87
Barbara Gallina 51
Gary Gilliland 299
Chris Hobbs 227
Alexei Iliasov 281
Yan Jia 15
Adam Johns 181
Pavan Venkatesh Kumar 87
Linas Laibinis 281
Peter Bernard Ladkin 149
Systra Scott Lister 281
John A McDermid 15, 227
Reuben McDonald 313
Paul McKernan 263
Catherine Menon 75
Elham Mirzaei 51
Paula Palade 115
Mike Parsons 123, 227
Zoë Porter 15
Austen Rainer 75
Alexander Romanovsky 281
Rachel Selfe 183
James Sharp 263
Mike Sleath 123
Dominic Taylor 281
Gerasimos Theotokatos 121
Harold Thimbleby 203
Carsten Thomas 51
Bernard Twomey 227
Dracos Vassalos 121
Gavin Wilsher 37

Printed in Great Britain
by Amazon